Time and Management

FRONTIERS IN MANAGEMENT SERIES
under the General Editorship of Dr. R. Alec Mackenzie

Flory and Mackenzie, The Credibility Gap in Management
Maher, New Perspectives in Job Enrichment
Webber, Time and Management

FRONTIERS IN MANAGEMENT SERIES

Time and Management

ROSS A. WEBBER

Wharton School of Finance and Commerce
University of Pennsylvania

VNR **Van Nostrand Reinhold Company**
New York Cincinnati Toronto London Melbourne

To
Mary C. Foradora
with whom time is always a pleasure

Van Nostrand Reinhold Company Regional Offices:
New York Cincinnati Chicago Millbrae Dallas

Van Nostrand Reinhold Company International Offices:
London Toronto Melbourne

Library of Congress Catalog Card Number: 75-181093

Manufactured in the United States of America

Published by Van Nostrand Reinhold Company
450 West 33rd Street, New York, N. Y. 10001

Published simultaneously in Canada by
Van Nostrand Reinhold Ltd.

15 14 13 12 11 10 9 8 7 6 5 4 3 2 1

Preface

The paradox of time and American management is this: we have been ambitious, conscientious people who believe that we must work diligently, utilizing every bit of present time, but we have been lazy and slovenly about the future. We have built the greatest economy and the strongest nation in history, yet a challenge remains—a challenge that requires us to value the present less and the future more. To enhance future time, managers must cease filling present time with activity: Not waste time now, but carve time out of the present for less-pressured work on long-range problems and possibilities.

The reader of this book, however, may be concerned mainly about practical advice on how he can manage his own time more effectively in a hectic present. Fine; the book does contain advice on this—although it is not intended primarily as a "how to manage time" text. College professors are frequently loath to offer such practical advice. We are reluctant not so much because we are ivory-towered hermits, but because all "how to" advice is partly wrong. How to books run the risk of implying that there is a right way of doing things, and that the author is an expert with the answers. But no questions about management can be answered with pat, exact replies. The era of books presenting absolute management rules and organization principles is past.

Nonetheless, the practitioner must act and some advice is necessary. Drawing on a wide variety of books and articles and the experience of many managers, the more applied chapters offer suggestions, not rules, and point out benefits and detriments of techniques for saving or managing time. Given your position and managerial philosophy, you must select what seems most useful.

A practical orientation toward time management is certainly legitimate, but not sufficient. The stakes are larger than a manager's control of his own time. Fostering organizational flexibility, creativity, and concern for the future involves more than just the efficient use of time today. Indeed, time efficiency in the short run may even be detrimental to time in the future. We must also discuss the philosophy of time and management. This requires some understanding of how managers think and behave; and how attitudes about time affect decision-making and planning.

Embodied in this book are four hopes: first, that you will gain some insight into yourself, your perspective on time, and your behavior at work; second, that you will learn some practical means to save time and manage it more effectively; third, that you will gain greater understanding about how time affects managerial behavior and decision-making; and fourth, that with such insight and understanding, you will develop your own philosophy of time and management that will reconcile the tensions of past, present, and future.

The book's organization is shaped by these hopes:

Chapters 1–4 deal with attitudes and behavior—your own, other managers, and those of various cultures.

Chapters 5–8 are concerned with applied matters, why time is wasted, and tactics for managing personal time.

Chapters 9 and 10 discuss how time affects and distorts organizational decisions, and compares individual and group decision-making.

Chapters 11–14 raise more philosophical issues about the time-binding role of management—integrating the past, guiding the present, and anticipating the future.

Chapter 15 concludes and summerizes in the form of analytical insights, practical tactics, and philosophical implications.

At the end of each chapter is a section of references and notes for readers who might wish more informatio.ı on specific topics.

In writing this book I have imposed on the time of many friends in business and academia. I am especially grateful to: Joseph Dooher and Alec Mackenzie for valuable comments on the manuscript; Jeremiah O'Connell for helpful insights; John Eavenson, Vince Boyer, and John Kemper for access to their organizations; Thomas Dunn, Alvar Elbing, Edward Halbert, and Toni O'Connell for contributing items; Eliot Chapple for access to his data; Leonard Sayles for early support of my interest in managerial behavior; my children (Sarah, Judith, Gregory, and Jennifer) for assistance; and Willis Winn and Edward Shils for time to write.

Those whose published work has particularly influenced this book include: Russell Ackoff, Sune Carlson, Peter Drucker, J. T. Fraser, Sebastian de Grazia, Robert Guest, Edward T. Hall, Paul Holden, Herman Kahn, Paul Lawrence, Harry Levinson, Rollo May, David McClelland, Elting Morison, Lewis Mumford, Reinhold Niebuhr, J. B. Priestley, and Leonard Sayles. I have tried not to distort their helpful views.

Finally, I am most grateful to my wife Mary Lou for critical advice and support in preparing, editing, and revising the manuscript.

In the introduction to his engrossing book *Man and Time,* the eminent British author J. B. Priestley dedicated the volume to those who like him are "time-haunted" men. Such an appellation is certainly appropriate for myself, for those people who have helped me, and for the readers of this book. Although intended for practicing managers, some old and fundamental issues about mankind are raised, The conscious and rational use of whatever time man could wrench from struggling to maintain life has been our great challenge and opportunity. It is no less so now in an era of apparently increased leisure. Our organizations must be better managed with a more serious view to the future if disaster is not to befall us.

R. A. W.

Cherry Hill, New Jersey

Contents

Time and Management

1

The Tyranny and Promise of Time

Yes? No? More? Less? Go? Stay? Now? Later? Slow? Quick? He? She? Black? White? Young? Old? New? Used? Lead? Lag? . . . Like Macbeth's battalions of troubles, a multitude of decisions press on managers—decisions that grow increasingly complex and time-consuming. And there's the rub; time does not grow with the expanding number, scope, and importance of these decisions. Indeed, it seems to shrink.

Long ago, the great physicist Isaac Newton suggested that time flows at a uniform, constant speed; every minute like any other. But everyone often imagines he was wrong. Time seems remarkably uneven in movement—now fast, now slow, depending on how one feels. A middle-aged man sits in his office watching his clock. The minute hand swings round and round. He thinks of all the other times and places when he stared at the clock with different attitudes. As a child lying in bed on Christmas Eve, minutes and hours seemed endless; but Christmas Day and its gaiety flashed by all too quickly. In school he wanted the wall clock to speed up so that precious free time would arrive. Outside was a ball, a car, or a girl—all desirable objects that the slow-moving timepiece seemed unaware of. But when he got there and with them, that merciless clock would appear to speed up!

In the military it was similar: counting the days, hours, and minutes until that most aptly named naval event called "liberty." He particularly remembers that first job after the service when as a staff specialist in a large corporation, time seemed to overflow. With other young professionals he sat by the telephone feeling all alone, if not actually so, awaiting calls for assistance and advice that came too infrequently. He had time, but like the ice cream consumed on an earlier job as a soda jerk, it had no meaning. With experience and performance, as he became a contributing member of the organization, that superfluity of time began to shrink. With promotions and age, time seems to fade away. (In fact, evidence exists that time actually goes faster as a man grows older; apparently this stems from a declining rate of chemical activity in the aging body).[1]

In school, military, and early jobs described, the hated *Now* would pass all too slowly to the desired *Then*. Now and Then were separate and distinct; none of the former spilled over into the latter. It was a world of simple dichotomy. Such is no longer true for the middle-aged manager. Now and Then are interrelated; what is not accomplished tends to fill up that future Then, so there is no escape from Now. As a result, instead of wishing those impersonal clock hands would speed up, as in high school English, at sea on a destroyer, and when first working, these days we more frequently wish they would slow down and stretch out the present. Talks with many managers indicate that these feelings are not unique; virtually all managers and professionals feel short of time.

Can nothing be done? Like receding hair, expanding girth, and increasing taxes, must the advancing man knuckle under to the seeming contraction of time with its corrupting influence on the quality of his decisions? Partly yes, partly no. Of course we are subject to time's tyranny, its passage, but we are also offered time's promise. The reconciliation of this paradox distinguishes man (and his unique managerial capacity) from all other creatures.

At the root of our problem with time and our desire to manage it is the human condition. In the Judeo-Christian tradition, God made man a creature both bound and free. With respect to time we are simultaneously aware of both our power and impotence. An ancient and gruesome image for time is that of an elderly man devouring his children. This grotesque creature was Chronos and the symbolism combined two attributes of time, the creative and the destructive.[2]

In this image the cannibalistic destructiveness clearly dominates the procreative. At a more realistic and personal level, a nineteenth-century symbol of time's tyranny showed a neglectful husband pursued by the spirit of his wife in the guise of a grandfather clock. With understanding, we can read the Latin words on an old sundial in Venice: "Horas non numero nisi serenas" ("I count only the hours that are serene"). By this criterion surely few clocks would show much movement in modern plants or offices—or homes for that matter.

On the more positive side it has been observed that "space-time" is the fundamental stuff of which the universe is constituted, and further, that it is the time part that offers promise. "Space provides room for being and time provides room for becoming . . . Space is the vehicle of determination and being, time is the vehicle of freedom and value."[3] Man's existence and the gift of time may be viewed as permission to partake in creativity of the world.

THE TYRANNY

It is difficult just to understand what time is even though it seems deceptively obvious. Most of us would concur with Saint Augustine's sentiment when he complained: "What then is time? If no one asks me, I know; if I want to explain it to a questioner, I do not know." Or with the modern philosopher Alfred North Whitehead who wrote, "It is impossible to meditate on time and the creative passage of nature without an overwhelming emotion at the limitation of human knowledge."[4] Yet, for all its complexity, it seems clear that time is irreversible, transitive, and successive. It moves on in a certain order that cannot be changed; it cannot be physically stopped, compressed, expanded, or reordered. No one can quick-freeze time like a side of beef or a misguided seeker of a cryogenic eternity. History, physics, and poetry attest to these aspects of time. Newton described time as absolute, flowing without reference to anything else. Thomas Carlyle advanced the same view more poetically: "The great mystery of *Time,* were there no other; the illimitable, silent, never-resting thing called Time, rolling, rushing on, swift, silent, like an all-embracing tide."

The Western world's emphasis on progress and growth is symbolized by the image of time as an arrow moving forward in a straight line from an ancient past to an unknown future. The imagery is particularly appropriate for the man of action posed before an information array and an approaching deadline. The decision is made, the deadline arrives, passes. Now old Omar's words take on meaning: "The Moving Finger writes: and, having writ,/Moves on: nor all your Piety nor Wit,/Shall lure it back to cancel half a Line,/nor all your Tears wash out a Word of it."

In short, men of action like businessmen and managers are particularly subject to time. Because time and attitudes toward life are closely related, the first citizens in Europe customarily to carry personal timepieces were businessmen in the seventeenth century.[5] The clock is the quintessential machine for the machine age—it introduces system, regularity, and impersonality to all who use it. Thus, the invention and popularization of mechanical clocks reflected and precipitated ideological change. "In the beginning," Sebastion de Grazia writes, "the clock exerted a strange almost morbid attraction as though it were ticking off life itself . . . Machines by now have manipulated everyone, their owners and tenders both, into living by the dictates of the clock. An ignorant visitor from a clockless land might wonder why we reject the tyranny of men while acquiescing to the tyranny of an idol."[6] Use of clocks in the monasteries of the Middle Ages gave regularity to human life at a personal level. Time became objective and impersonal—relative to the movement of the mechanical timepiece rather than to the seasons or natural rhythms that had dominated in the past. Timekeeping became time serving; time serving expanded human consciousness; and Now replaced Eternity as a primary concern.

Some scholars have suggested that such a paced, orderly, punctual life was not natural, that such order and speed reflected a reorientation of man's perspectives away from the older view of time as unlimited, cheap, and reoccuring to those attitudes that characterize us today—that time is limited, precious, and one-way.[7] In its earliest days, carrying a timepiece was more than just a symbol of status; it was a mark of wisdom and virtue. Today we say that using time correctly should reflect these worthy attributes.[8] To be prudent and concerned about time is a trait we strive to teach our children.[9] Perhaps Marshall McLuhan is correct: "The Eskimo," he expounds, "is a servomechanism of his kayak, the cowboy of his horse, the businessman of his clock."[10] For many of us the lines of Michel Quoist are all too true:

Good-by Sir, excuse me, I haven't time.
I'll come back, I can't wait, I haven't time.
I must end this letter—I haven't time.
I'd love to help you, but I haven't time.
I can't accept, having no time.
I can't think, I can't read, I'm swamped, I haven't time.
I'd like to pray, but I haven't time.[11]

THE PROMISE

But if passing time cannot be slowed and the past not restored can managers do nothing but make their decisions as they come? Must they always be passive respondents to the external forces that converge on them demanding action?[12] Of course not. Although no manager can be entirely free, they can and do *manage* at least some of their time. This book clearly hopes to assist in this endeavor.

Indeed, businessmen historically have been quite creative in attacking time. Corporations, which grew primarily from a need to raise capital and share risk, are an example of an effective instrument to overcome time. Time is not the same for the corporation as for the entrepreneur; it is as marble to sand castles. If effectively structured and administered, the company can outlive the founder. This drive to master time has been a significant factor in the growth of social institutions.

This book, however, is concerned with individual managers and their life on the job. It will survey various suggestions that have been advanced about how managers can manage personal time more effectively. To do this, however, managers must have some insight into why they allocate their time as they do and what factors influence their behavior. For only by working on these factors can they change the time patterns of their jobs.

We do not want to sound too mechanistic or overly rationalistic. We are not attempting to develop a PERT chart on how managers should spend their time. Much of the available advice of time management suffers from just such bias. It pictures a cool and rational executive allocating his time in advance of events according to some objective criteria related to organizational goals. Everything is managed. But this is impossible. To foster such an image is misleading and unfair to harried executives. Realistic talk is needed, not more advice telling managers to be something they cannot. They are not able to plan and control everything—even their own time.

An overly rational approach tends to blame others for a manager's inability to control his time. Unlike the private physician who manipulates his patients to suit his convenience, managers mainly respond to others who determine how their time is utilized.[13] Complaints about this situation often seem to offer a wistful wish: "Wouldn't management be easier and better if there were no other people except me?" But of course this is nonsense. This view might be related to a poll of college seniors that indicated overwhelming preference for entering independent professions rather than business or government.[14] Partly they are trying "to do something for people" (although their rejection of government careers seems contradictory); partly they are trying to "avoid office politics." Nonetheless, their strongest reason seems to be freedom from others' demands, control of their own time—an attractive, but not altogether praiseworthy, motivation. A manager has many roles to play. One of these is as a response agent, subject to the control of others. An effective manager sometimes must fill this part.

Perhaps the reason why so much of the literature and advice to practicing managers seems so overly rational lies in the different relationships between authors and their jobs and managers and theirs.[15] The intellectual observer usually has substantial control over his working schedule; he has nobody to supervise besides himself. In addition, when he thinks about management, it is alone in his office, or on the beach, or wherever the scholar has the luxury to think and write. He tends to feel that with good will and intelligence all sorts of affairs can be planned and implemented. He lives in a different world than the manager for whom it is just enough to get through the working day in one piece. From where the manager sits the world is more capricious, less subject to control, more a series of crises. His overriding thought frequently is to get through each problem in good enough shape to face the next. "Life," as Mark Twain is reputed to have once observed, "is just one damned thing after another." We shall see that total management of a manager's time is not possible, nor even desirable. Some time cannot and should not be managed. In short, at times managers must drift.

At a modest level this book endeavors to assist the manager and prospective manager in reconciling the tyranny and promise of time. No panaceas will offered, no miracles revealed. Time's illimitable flow precludes such possibilities. Nonetheless, in an area so vital, even small advances are large in importance. As the manager's use of time is discussed, perhaps a sense of familiarity or shock of recognition will be aroused. Good. Analysis may make explicit many things taken for granted. More important, reading, talking, and thinking about these matters may free us a little from time's restraints.

REFERENCES

1. On the shrinkage of subjective time with aging, "possibly as a result of the gradual slowing down of metabolic processes, and hence retardation in the rate of healing," see John Cohen, "Subjective Time," in J. T. Fraser, ed., *The Voices of Time: A Cooperative Survey of Man's Views of Time as Expressed by the Sciences and by the Humanities* (George Braziller, New York, 1966), p. 267.
2. The various symbols and images for time are discussed by Joost A. M. Meerlo in "The Time Sense in Psychiatry," in Fraser, *op. cit.;* also by J. B. Priestley in *Man and Time* (Aldus Books, London, 1964).
3. The concept of "space-time" is from Samuel Alexander and is cited by Cornelius Benjamin in "Ideas of Time in the History of Philosophy," in Fraser, *op. cit.* p. 25.
4. Whitehead is quoted by Benjamin in Fraser, *op. cit.* p. 8.
5. On the development of the clock and its relationship to ideology, see Lewis Mumford, *Technics and Civilization* (Harcourt, Brace & Co., New York, 1934); on the businessman's use of the clock and personal timepieces, see Miriam Beard, *A History of the Businessman* (University of Michigan, Ann Arbor, 1961).
6. The quotation from Sebastian de Grazia is in his monumental, *Of Time, Work and Leisure* (Twentieth Century Fund, New York, 1962), p. 307. (© 1962 by The Twentieth Century Fund.)
7. The suggestion that a paced, orderly clock-directed life is unnatural is by Mumford. In contrast, an anthropologist argues that nothing is natural to man; man and his culture are so plastic that "There is no natural man," he asserts; and we cannot "talk about artificial manners . . . because all manners are artificial." See Marston Bates, *Gluttons and Libertines: Human Problems of Being Natural* (Random House, New York, 1968).
8. A psychiatrist writes how time is experienced as a great censor of life "which keeps us caught in restrictions and coerced moral behavior. One of my patients expressed it sarcastically: 'Time is civilization,' by which she meant that strict scheduling and meticulousness were the great weapons used by adults to force youngsters into submissiveness and servility." (Joost A. M. Meerlo, "The Time Sense in Psychiatry," in Fraser, *op. cit.*, p. 249).

9. To be concerned about time is also a trait managers wish that new employees from the urban ghetto would learn. One survey of experience with such employees observes, "The most serious problem after training continued to be lateness and absenteeism. . . . Those of us who have lived by the clock from our earliest days take the alarm clock for granted as essential in getting us to work on time. But in many inner-city households this appliance is not part of the standard equipment, for it has never been needed. The counselor should see to it that every new hire has one. At Westinghouse/East Pittsburgh, after the first paychecks have been distributed, the counselor accompanies the new employee to the nearest store to make sure he gets an alarm clock." Leonard Nadler, "Helping the Hard-core Adjust to the World of Work," *Harvard Business Review,* Vol. 48, No. 2 (March-April 1970), p. 123.

The pressure to utilize time is not just a European phenomenon (although perhaps it had more lasting impact there). Six centuries of mechanical clocks in China preceded the appearance of such clocks in the European west. Men of action had similar relations to time everywhere. A 1500-year-old Chinese text has an interesting section entitled Hsi Shih (Sparing of Time): "The worthies of old, wishing to spread benevolence and righteousness in that world, were always struggling against time. They set no value on foot lengths of jade, but a tenth of an inch of shadow was as precious as pearls to them. Thus it was that Yu the Great raced with time to finish his work and paid no attention to the enquiries of Nanjung. Thus it was that Tao Chung never stopped walking till the soles of his feet were as hard as iron. Confucius grudged every moment lost from revealing, and Mo Ti was up and about again before his bed had time to get warm."

What hard-charging American executive could do more? The book is *Hsin Lun (New Discoures)* attributed to Liu Hsieh (ca. 550 A.D.); it is cited by Joseph Needham, "Time and Knowledge in China and the West," in Fraser, *op. cit.,* p. 106.

10. Marshall McLuhan is quoted in an interview in *Playboy,* March 1969, p. 53.

11. Michael Quoist is quoted on the opening page of Staffan B. Linder's intriguing book, *The Harried Leisure Class* (Columbia University Press, New York, 1970). Linder writes: "We have always expected one of the beneficial results of economic affluence to be a tranquil and harmonious manner of life, a life in Arcadia. However, the pace is quickening, and our lives in fact are becoming steadily more hectic." People should have more time for leisure, but instead they have increasingly less time. They are overwhelmed by the number of goods and services available—there are many more things to do, but the amount of time remains unchanged. Rather than enjoying them, people are becoming slaves to the possessions and services that compete for their leisure hours. Linder argues that "One may possibly *buy* more of everything, but one cannot conceivably *do* more of everything." pp. 1-2. The lines by Michael Quoist appear in *Prayers,* by Michael Quoist, © Sheed and Ward Inc., 1963, and are reprinted here by permission of the publisher.

Part of the problem of harried leisure time may lie in the commercialization and streamlining of leisure time so that pressure is turned on and we begin to feel obliged to do such and such, and it all ceases to be leisurely. An editorial in *The MBA* of April 1971 observes that "one of the most appalling acts of Congress in recent years was the wholesale reshuffling of once-immoveable national feasts so as to conform to the fantasies of a group of motel owners and tour bus operators. The net effect has been to destroy the delightful randomness of the calendar, depersonalizing, dehumanizing yet another area of national life. Remember when George Washington's birthday fell on a Wednesday? The luxury of a day-off in mid-week? Nevermore." p. 4.

12. Some of us do become dominated by time and its monitor, our clock. Consider the following observations in a recent article on the poet W. H. Auden: "With all the freedom in the world...Auden chains himself to time. Its beat dictates his life. He looks at his wristwatch and says, 'I'm a slave to it. I shouldn't know what to do unless it told me.' And he checks his watch over and again, as if requesting a command or confirming an order. Eating, drinking, writing, shopping, crossword puzzles, even the mailman's arrival—all are timed to the minute with accompanying routines. . . . And Auden will go on structuring time like Eric Berne's people playing games because the more structured the time the less threatening its passage." [Webster Schott, "Autumn for the Age of Anxiety," *Life,* Vol. 68, No. 3 (January 30, 1970) p. 53. © 1970 Time Inc.]

13. Peter Drucker's very helpful views of time management are expressed in his *The Effective Executive* (Harper & Row, New York, 1967), p. 10. He writes: "Indeed, unless executives work at becoming effective, the realities of their situation will push them into futility.

"Take a quick look at the realities of a knowledge worker *outside* an organization to see the problem. A physician has by and large no problem of effectiveness. The patient who walks into his office brings with him everything to make the physician's knowledge effective. During the time he is with the patient, the doctor can, as a rule, devote himself to the patient. He can keep interruptions to a minimum. The contribution the physician is expected to make is clear. What is important, and what is not, is determined by whatever ails the patient. The patient's complaints establish the doctor's priorities. And the goal, the objective, is given: It is to restore the patient to health or at least to make him more comfortable. Physicians are not noted for their capacity to organize themselves and their work. But few of them have much trouble being effective.

"The executive in organization is in an entirely different position. In his situation there are four major realities over which he has essentially no control. Every one of them is built into organization and into the executive's day and work. He has no choice but to 'cooperate with the inevitable.' But every one of these realities exerts pressure toward nonresults and nonperformance."

14. The poll referred to was by Louis Harris Associates for *Newsweek* magazine (May 2, 1966; reprinted by permission). The results for 800 college seniors across the country are summarized as follows:

Where to Go to Work?

"Asked to name the career they would like best, college seniors rank business far down the list. But many are resigned to winding up in business."

	Free Choice	Think They'll Enter
Profession	55%	33%
Teaching	24	40
Business	12	21
Government	9	6

15. These views on the gap between intellectuals and practicing administrators are from recollections of an eminent philosopher about his service as an Assistant Secretary of State in the Johnson administration. See Charles Frankel, *High on Foggy Bottom* (Harper & Row, New York, 1969).

2

Perspectives on Time

Time is not the same for everyone. Rather like the inkblot of psychological fame, people project certain aspects of their cultures and of themselves onto their conceptions of time. Time seems to have special meaning for those attracted to managerial careers. To explore the interrelation between personality, culture, and time, consider the following tests of time perspectives.

TESTS OF PERSPECTIVES ON TIME

Test One: Time Metaphors

Which image of time do you prefer?[1]

(1) A quiet, motionless ocean	______
(2) A galloping horseman	______

Test Two: Time Description

What two words best describe the idea of time for you?

sharp	active	empty	soothing	tense	sad	clear	cold	deep

Test Three: Past Events

Without consulting any references or persons, when did the following events occur?[2]

	your date	*actual*	*your estimate is correct*	*over*	*late*
(1) Outbreak of the Korean War	___	___	___	___	___
(2) Cuban Missile Crisis	___	___	___	___	___
(3) Supreme Court decision on school segregation	___	___	___	___	___
(4) Overthrow of Khrushchev	___	___	___	___	___
(5) Assassination of Martin Luther King	___	___	___	___	___

Now fill in the actual dates (they may be found on page 14). Determine if you were correct, overestimated, or underestimated the time elapsed since each event.

Test Four: Checking Your Watch

Try a simple experiment that may suggest how your perspective on time is translated into behavior. Check your wristwatch or bedside alarm clock for accuracy. The radio or telephone may be used to determine the exact time; compare this with your timepiece.[3]

Your watch/clock time	______		
Actual time	______		
Your watch/clock is	______ fast;	______ slow	

Such paper and pencil questions are easy; the following test is more difficult, but it may help you to *feel* time.

Test Five: Perceiving Time

You will need a helper for this one. Find a quiet room where you will not be disturbed; ensure that you will not be interrupted. Remove all timepieces. No reading matter should be available, nor any work to perform. Quiet music can be playing, but not a program that announces the time. No television. Before closeting yourself in the room, have a member of your family or friend agree to call you after a period of time *they* determine (as a guide, 10 to 40 minutes). Now go sit, think, or dream. In the vernacular of the day, tune out, turn off, and drop out—for just a short time at least without pharamacological or artificial assistance.

When your collaborator calls, before he indicates the time, estimate how much time passed while you were sitting.[4]

your estimate	______mins	your estimate is
actual time	______mins	______correct, ______over, ______under

WHAT THE TESTS MEAN

These exercises have been given to many people particularly to explore the "need for achievement." Persons high in this need tend to share certain perspectives on time and to gravitate toward achievement-oriented careers such as management. Why? To explain we must examine the nature of achievement need and discuss its implications for attitudes toward time.[5]

The Need for Achievement

"To achieve: to carry to a termination, to bring into a perfected state or a successful conclusion; to obtain or gain as a result of exertion." The need for achievement drives men to activities that lend this satisfaction. In stories the need is expressed by the hero who overcomes great obstacles and arrives at a distant goal through persistent effort. Note that emphasis is not on the goods or honors that are obtained through successful achievement. Rather, the need is concerned more with satisfaction through the *process* of achieving. For example, great achievements may bring social prestige, power, security, perhaps even love. As satisfying as such rewards are, however, they do not reflect the achievement need itself. Men with high need for achievement may enjoy status and money as much as other people, but the need also includes intrinsic satisfaction with (a) performing a task well and meeting high standards, (b) trying hard to overcome difficult (but not impossible) obstacles, (c) trying different ways, novel and creative solutions. Thus, the high achiever commits himself emotionally to tasks which he considers worthwhile and difficult

So the achievement-oriented man is attracted to activities of some challenge, difficulty, and risk (but not long shots or sure things), some novelty and variety where he can personally affect the outcome and obtain gratifying feedback on results. He is attracted to management. In a number of countries, managers have been found to be higher as to achievement need than independent professionals. In the United States, middle managers seem to have higher achievement needs than a random sample of similarly aged college graduates. Finally, managers with higher achievement needs tend to earn more money. As one corporation executive vice-president puts it: "If there is anything you need, it is a sense of urgency. You must develop the ability to be impatient at times. The man who will be successful as a manager must develop a recognition of the importance of time."

So far everything suggests that to have a high need for achievement is a good thing. Jobs will be exciting and the money high. But of course this is not inevitable; other factors (such as ability) also count; and besides, certain costs are associated with achievement need. One of these costs is that the high achiever may be especially "time haunted."

Time and the Need for Achievement

Movement, direction, and value characterize the high achiever's orientation toward time. On test one he chooses swift, moving images for time like "a galloping horseman," (or "a fleeing thief," or "bird in flight") rather than "a quiet, motionless ocean."

Complementing this dynamic perspective, high achievers tend to describe time in terms of "clear, sharp, active, tense, fast" rather than "empty, soothing, sad, cold, deep," which are more likely to be picked by persons lower in need for achievement.

It appears that men with high achievement needs are more concerned about time's passage. They tend to recall the past events in test three as nearer the present—they underestimate the time that has passed. The actual dates:

(1) Outbreak of Korean War, June 25, 1950.

(2) Cuban missile crisis, October 16–29, 1962.

(3) Supreme Court decision on school segregation, May 17, 1954.
(4) Overthrow of Krushchev, October 1964.
(5) Assassination of Martin Luther King, April 5, 1968.

When sitting alone in a room as in test five, high achievers tend to be either more accurate or to overestimate how much time has passed. Those lower in achievement need tend to underestimate—perhaps because they are less active normally, less impatient, less concerned about lost time, and because they become passively absorbed. Other research indicates that those higher in achievement need tend to feel: (a) more annoyance when "your watch has stopped or is not running properly," (b) guilty "if you sleep late in the morning," (c) certain that "you waste time or spend it uselessly," and (d) anxious "when you are not certain of the time."[6]

But how could attitudes about time affect a person's watch? So might be the reaction to test four. Yet research suggests that high achievers often have watches that are fast! Although not quite so strikingly, those with lower achievement need often have watches that are slow. Unless one has enormous faith in the power of human brain waves to influence mechanical parts, this seems remarkable. Yet, it is not. Some people simply set their watches a few minutes ahead in an effort to create time, a little game to delude themselves into getting ahead of tyrannical time. They are, so to speak, always a little ahead of themselves. It doesn't always work, however, because high achievers still tend to be late to appointments—perhaps because they get involved in too many activities.[7]

The future also looms more importantly to those higher in achievement need. They tend to have longer time perspectives in imaginary stories that they write. Students higher in achievement need have better recall twenty-four hours after receiving an assignment due eight weeks later. Students lower in achievement need tend to forget about such a long distance assignment and concentrate on those nearer. Semantics scholars even report greater use of anticipatory tenses in the language of more achievement-oriented cultures.

Achievement orientation is also associated with ability to delay gratification. Given a choice between a small reward now or a larger reward later, most youngsters would probably pick the former, but those higher in achievement need are more likely to choose the future prize. Nonetheless, high achievers tend to be impatient with rewards that follow too long after the performance. They want concrete results indicating how they have done and are frustrated if these are too long delayed. Consider the high achiever as a teacher. He may be distressed because it is so difficult to determine how he he is doing; course examinations may indicate only how well his students memorize, not their learning. More valid results are indicated by the students' lives after graduation, but the teacher probably will never know about this. Even if one of his boys makes it to the White House, given political realities, he is not likely to take a mentor's satisfaction in his former pupil's performance.

Drawing on Greek mythology, David McClelland describes the winged-footed Hermes/Mercury as a symbol of achievement (and of businessmen): He is "in effect a dynamo of energy—always on the move, in a hurry. He doesn't waste a moment's time, but starts out into the world to get ahead the moment he is born. He makes his technological innovations and pursues his career in a spirit of restless energy in which there is much motion and less waste of time . . . the picture of a man hurrying forward with his attention focussed on the more distant future so present time seems to be slipping past him rapidly."[8]

TIME AND AMERICAN CULTURE

We Americans are less than Greek Gods, but we inhabit a nation and age where the Hermes-like concern for time has been widespread.[9] That all peoples do not share the same view of time is implicit in actual complaints voiced by United States managers doing business abroad: (1) Why are foreign managers so late for business appointments? (2) When Americans and foreigners do meet, why are the latter so slow in getting down to business?[10]

Tardiness seems to reflect the individual's and society's conception of time. Is it a scarce commodity to be saved and used carefully? Or is it an endless resource to be consumed carelessly? To picture the matter schematically, time may be seen either as a circle or a straight line. The circle suggests repetition and continuity for time is measured by natural events—the sun's movement, the moon's phases, the seasons, birth and death. Measurement is gross; minutes and hours have no meaning. If the present is wasted, it will come around again. Today will return tomorrow for life and time are endless repetition. With such a philosophy life appears unhurried, even serene.

How different is the linear view! Yesterday is gone forever, today is here but a moment, and tomorrow depends on what we do now. Time is less associated with natural phenomena, more with the artificial movement of a mechanical timepiece (which is only circular for convenience; many early clocks were linear and the newest display only numerals). Time becomes measured and referred to as precise points, not ranges. Life appears cluttered and hectic as people respond to time's demands.

These attitudes are reflected in religion. Thus for some Oriental groups, life is a long, repetitive series of reincarnations in which lives are lived under various guises depending on the spiritual worthiness of the past life—a crab becomes the new form for the vainglorious, an eagle for the prudent, and so on. Salvation is not the achievement of a specific life form, however, it is release from time's circle. It is escape into a dimensionless Nirvana where time is irrelevant. Occidental salvation tends to be different. Whether the Judeo-Christian or Marxist versions, time is going somewhere—not to nothingness, but to a utopia here or in heaven. Time and life thus share a mission, have purpose, and should be lived correctly.[11]

So most Americans have been exhorted by moralists to make the most of time: "A waste of time is the worst of sins," "Idle hands are the Devil's workshop," and so on.[12] Similarly, after Lenin took power in Russia in 1918, he began to preach that time was to be preserved and used effectively. The superfluity of time that the peasants enjoyed (or found monotonous) on the steppes should not exist for citizens in the new workers' state. Here is part of a leaflet distributed by the Soviet Time League whose members were obliged to protest against and report every waste of time they encountered.

> Measure your time, control it!
> Do everything on time! Exactly on the minute!
> Save time, make time count, work fast![13]

Not very poetic, but not unlike the philosophy of many American managers.

The American manager (and probably the contemporary Soviet) may be correct when he complains about foreign managers being late for appointments. For them, time just may be less precious—or other activities more important. And as the old adage goes, concern with time utilization and practice in using it simply may make the United States manager more skilled in planning to be on time, arranging transportation, and all the other requirements of a time-conscious society. As a result, in Saudi Arabia one may be on time anywhere within a three-hour span. In New York City the acceptable span may be five minutes (although growing transportation problems in New York may be broadening this range).

Waiting for someone in our time-conscious society is one of the sharpest symbols of inferiority. Hurt pride and revengeful feelings frequently creep into the waiter's mind under even the best circumstances. Unless a manager is intentionally executing a Machiavellian maneuver designed to impress a subordinate that he *is* inferior, he should avoid having people wait.[14]

Another insight into American attitudes toward work and time is revealed by what executives frequently say about vacations: one often apologizes for taking a vacation by explaining it is only in order to recharge his batteries—he justifies rest or play mostly in terms of returning to do a better job. A former President of Du Pont states "My wife and I have a place in Bermuda. We used it as a rest cure. I decided that I couldn't work at being president of the company for eleven months at a time and then take a month off. I'd get irritable and unhappy during those eleven months. So we began taking short trips to Bermuda—three or four times a year. . . . When I'd come back from one of these trips, I'd feel better and make decisions more easily."[15] By contrast, many Europeans seem to enjoy their vacations as a pleasure in its own right. They may in fact be more likely to work for the express purpose of being able to afford to play in a grander style.

America today seems to be reexaming this utilitarian view of time. At least some argue that we have been automatons to the clock, that self-actualization can only be found in withdrawal from the system. Yet, the depth of our commitment to using time is reflected in the emotion of the disavowals. Thus if a young American wants to loaf for an extended period of time, he tends to justify it to this effect: "The system is corrupt, I reject it. Its values are not my values. To hell with these puritanical, obsolete concepts." If he is more guilty and articulate, he may go on: "All of those robots who are working have sold out to the soulless, inhumane system with its obsolete and grubby, machine-based, materialistic values, and its empty goals of the Bitch-Goddess of monetary success. By refusing to be drawn in, I can preserve my humanity, individuality, integrity, dignity, and physical health, as well as my spontaneity, the freshness of my perceptions, the openness of my relating and my capacity to love."[16] Only after this disavowal can the young American flee work and begin hiking in the Big Sur country. A young African manager pointed out to me that this was nonsense; all that is necessary is to know how pleasant it is to "lie back on the grass under a shady tree, chew on a piece of grass, stare at the sky and watch occasional flocks of birds fly by."

In spite of much current rhetoric then, time retains its importance for most Americans. Whether hippie or establishment, we are concerned about its utilization. As we have seen, many businessmen and managers are especially driven by their need for achievement, by desire to apply their time to challenging problems, to offer novel solutions, and to obtain gratifying feedback on what they have accomplished within a reasonable time. They are told to use time. Most heed the advice; they use time in activity. Yet, they still have the problem of judging when this activity-filled time is being used rightly or wrongly.[17]

REFERENCES

1. The time metaphors and descriptive words are from R. H. Knapp and J. T. Garbutt, "Time Imagery and the Achievement Motive," *Journal of Personality,* Vol. 26 (1958), pp. 426–434.
2. Perception of past events is reported by H. B. Green and R. H. Knapp in "Time Judgement, Aesthetic Preference, and Need for Achievement," *Journal of Abnormal Social Psychology,* Vol. 58 (1959), pp. 140–142.
3. The relationship between watch time and achievement need is reported by David McClelland, *The Achieving Society* [1] (Van Nostrand Reinhold Co., New York, 1961), p. 327.
4. Perception of passing time is reported in Knapp and Green, "The Judgement of Music-filled Intervals and *n*-Achievement," *Journal of Social Psychology,* Vol. 54 (1961), pp. 263–267.
5. The seminal books on achievement need are D. C. McClelland, J. W. Clark, R. A. Lowell, and E. L. Lowell, *The Achievement Motive* (Appleton-Century-Crofts, New York, 1953); and McClelland, op. cit. A popular report of McClelland's research is his "Business Drive and National Achievement" [2], *Harvard Business Review,* July-August 1962, pp. 92–112.
6. The consideration of guilt and discomfort feelings in relation to time is in Knapp, "Attitudes Toward Time and Aesthetic Choice," *Journal of Social Psychology,* Vol. 56 (1962), pp. 79–87.
7. Attitudes toward time seem to be related to those hoary Puritan virtues of neatness and cleanliness. A psychiatrist writes: "The English word *time* has manifold meanings. It originates from a root *ti*—to stretch—and is in this way directly related to span and duration....Another linguistic link is to tide, tidiness and good tidings. Psychoanalytically, we

are very much aware of the relation between time, cleanliness, and orderliness." Joost A. M. Meerlo in "The Time Sense in Psychiatry," in J. T. Fraser, *The Voices of Time*, p. 247.

8. Information on perspectives on the future and gratification delay are drawn from McClelland, op. cit. [1], pp. 324–329, the description of Hermes, p. 303.

9. Sebastian de Grazia's *Of Time, Work and Leisure* contains the following insightful passage on time and work in America (pp. 305–311): "Time today is valuable. The clock's presence everywhere, and its tie to the factory with its relatively unskilled work, soon gave rise to the idea that one was selling time as well as, or rather than, skill. The lightening of toil and simplifying of tastes brought about by machines gave a related impression: that one was selling time rather than labor. The 'hourly rate' and the 'piece rate' express these notions. So time begins to be money, and, like money, a valuable, tangible commodity, to be saved, spent, earned, and counted....No other nation by now is as precise in its time sense nor so time conscious as the United States. Americans generally are aware that time runs by steadily and is being used up evenly, minute after minute, hour after hour, day after day—inexorable, impersonal, universal time....The American office schedule is tight and sacred. 'I'll see you at 4:10, then' is a sentence that would have been comprehensible to no other civilization this earth has seen. Violators of the schedule are punished. If you are not on time for appointments you will be regarded as an irresponsible person. If a man is kept waiting in the outer office for ten or fifeteen minutes, careful apologies are necessary. In some countries, in the Ottoman Empire tradition, a man can be kept waiting without offense for an hour or an hour and a half.... The clock then with its precise units breaks the day into equal parts that by conscious decision are to be filled with worthy activities. A man may want to loaf his time away, yes, but loafing is wasted time, and time shouldn't be wasted. It is valuable and scarce. One has only 24 hours of it a day." (© 1962 by The Twentieth Century Fund.)

10. These questions about American and foreign punctuality are discussed by James A. Lee in "Cultural Analysis in Overseas Operations," *Harvard Business Review* (March-April 1966), pp. 106–114.

11. Joseph Needham maintains that the Hebrews were the first Westerners to give value to time. For them, time went one way; there were no repeat performances. This conciousness directly contributed to the rise of modern technology and science. He writes: "It may be what recurrence theories (of time) really sapped was the psychology of continuous, cumulative, never-completed natural knowledge." If knowledge is to disappear, why bother? "Must not the early reformers and merchants alike have believed in the possibility of revolutionary, decisive, and irreversible transformation of society?" ("Time and Knowledge in China and the West," in Fraser, *op. cit.*, p. 131).

12. That Americans have been uniquely subject to pressures about time is illustrated by a Massachusetts colony statute of 1633: "No person, hawseholder or other, shall spend his time idely or unproffably, under pain of such punishment as the court shall thinke meet to inflicte."

13. The Soviet pamphlet was distributed in 1923. It is quoted in Reinhard Bendix, *Work and Authority in Industry* (John Wiley, New York, 1956), p. 208.

14. On waiting, Meerlo writes, "Primitive man did not and does not mind waiting; but in our culture waiting is often interpreted as an act of hostility....waiting for somebody has become in our society of schedule the symbol of either rejection or being less powerful. Hurt pride, depression and worry may fill the interval of waiting with dreary anticipation of revenge." (*op. cit.*, p. 249). An interesting example of the symbolism in waiting is provided by a newspaper account of a visit of Senator Edward M. Kennedy to West Germany: " 'Your Teddy Kennedy is a rube,' a senior protocol officer of the Bonn Government recently said to an American acquaintance . . . What annoyed the protocol official and nearly everyone else who had anything to do with the Kennedys here was their habit of showing up late for every appointment. The Germans in this century have not enjoyed a reputation abroad for good manners, but one point of etiquette that they carefully observe is punctuality. So it was that Kennedy tardiness made a bad impression not only on their official hosts but also on the German press . . . An American diplomat attempting to soothe the Germans with the explanation that tardiness was 'customary and planned' by the Kennedys as a means of increasing the anticipation of their audiences." David Binder, *Herald Tribune,* Paris, April 23, 1971. (© 1971 by The New York Times Company. Reprinted by permission.)

15. Crawford Greenwalt, quoted in *Lessons of Leadership,* edited by the editors of *Nation's Business* (Doubleday & Co., New York, 1966), p. 266.

16. These observations on youth's attitude toward time are from Herman Kahn and Anthony J. Wiener, *The Year 2000* (Macmillan, New York, 1967), p. 214. (Copyright © 1970 by The Hudson Institute, Inc.)

17. On the right use of time, Priestley reminds us: "There are dedicated savers of Time who might as well be social insects: Little good and much harm have come out of their saving. There are apparent squanderers leading scandalous lives, who have suddenly brightened the glory of our species. But, even they, in the end have used their time." (*Men and Time,* Aldus Books, London, 1964, p. 60.) After this discussion of man and time, we should observe that man's sense of time is not necessarily superior to all other creatures. "For it is conceivable," writes John Cohen, "that our reliance on watches and other artificial aids has led to an atrophy of our sense of time. And it is salutary to reflect that the minute beach crab, with barely 1 millimeter of nerve substance, is able to 'calculate' the time of day, the course of an hour, as well as the angle formed by its bodily axis and the position of the sun." Would that we could keep as close track on our progress through day's activities. ("Subjective Time," in Fraser, *op. cit.*, p. 257.)

3

Present Time: Typical Managerial Behavior

Ever receive a phone call in the middle of the night? In a mixture of sleepy confusion and awakening panic one bumbles around for the receiver. Such a call is apt to be taken as a matter of life and death; it is a particularly vivid example of how time talks. The timing and duration of a communication convey a message just as meaningful as the words said. When a consultant conducts early interviews with a client's managers regarding a management development program, the men who keep him waiting and who never seem to have as much time as planned are suggesting that they are not pleased about the program—a point doubly made if the managers remain behind their desks neither rising to greet the consultant nor to talk without the protective distance of the desk between them.[1]

How managers utilize their time tells much about them and their jobs. In this chapter we shall examine typical managerial behavior. This is not necessarily a description of how any one manager spends his time. Rather, the data will reflect the average of one heterogeneous sample of managers.

HOURS OF WORK

One fact seems to characterize almost all managers: they work long hours. According to data from the household surveys of the Bureau of Labor Statistics, managers put in more time than any other nonfarm occupational group. The list runs: (1) managers, officials, and proprietors 47.0 hours per week, (2) craftsmen and foremen 40.9, (3) professional and technical 39.6, (4) operatives 39.6, (5) sales workers 36.6, (6) clerical workers 35.6, (7) laborers 34.4, and (8) private household workers 22.0.[2] Other research indicates that the average manager spends approximately 43 hours per week at the office. On top of this, however, must be added seven hours per week at home doing paper work and business reading. Another five hours on business entertaining, and perhaps five hours on commuting (which is probably low) round out the picture—a grand total of some 60 hours per week.[3]

In addition to managers performing their company duties, some organizations have also encouraged and even expected executives to engage in community activities. Indeed, such participation is sometimes the best way for an ambitious young manager to demonstrate his abilities and bring himself to the attention of higher management. For executives, accepting posts in community organizations has often just gone along with the job. In recent years, however, these outside commitments have been making increasing demands on managers. A survey of corporations observes that increased time pressures from unanticipated growth in the workload of external activities has been one of the salient features of the past twenty-five years.[4] It is becoming clear that community activities cannot be something that a manager performs exclusively on his own time; these are expanding too rapidly. In the future, such activity will have to be incorporated into a manager's organizational position.

A growth in home demands also affects management. Today's average middle-class manager (if he is married) may have less free time than his counterpart in 1900, when the workweek was supposedly much longer; some of the vaunted reduction of the housewife's working hours has occurred by transferring labor to her husband. It is not so much the old issue of sacrificing family for career. All ambitious men do this to some degree (usually with some encouragement from their wives). The emerging problem is caused by essential home duties that formerly were performed by household workers in an earlier era when this was feasible. Both Presidents Kennedy and Johnson encountered some difficulties with their subordinates in this area. Insensitive to these off-the-job home demands because of personal wealth, both Presidents were unmindful of the demands on their less affluent subordinates. Of necessity, they required long working hours, but seemed to assume that their assistants were completly free for rest and relaxation when not on duty. Given predictions of continued decline in the quality of public services, the near disappearance of domestic help, and growing demands for off-the-job study in order to keep up with expanding knowledge, managers will have increasing difficulties in reconciling job and home demands on their time.[5]

TYPICAL COMMUNICATIONS

Years ago Elton Mayo, the great pioneer of the behavioral study of management, suggested that "social studies should begin with careful observation of communications." Virtually all empirical studies conclude that much of management's time is consumed by communications—as much as 60 to 90 percent on verbal communications, perhaps 25 to 50 percent in conferences, with written communications taking relatively little time. Our research supports these observations.[6]

Incoming Communications

Incoming communications are those that others initiate to the manager. They are not just one-way; on some the receiving manager just reads or listens, but with more he becomes involved in a two-way conversation.

The typical manager thus spends almost half (42 percent) of his time involved in communications that others have initiated. He should be able to sit, listen, discuss and/or read for substantial time. As might be expected, communications from subordinates consume the most time. The next most come from diagonal and lateral associates outside his direct chain of command. This includes staff personnel and managers of other departments. The typical manager spends more time on communications from his subordinates and associates than from his direct supervisors.

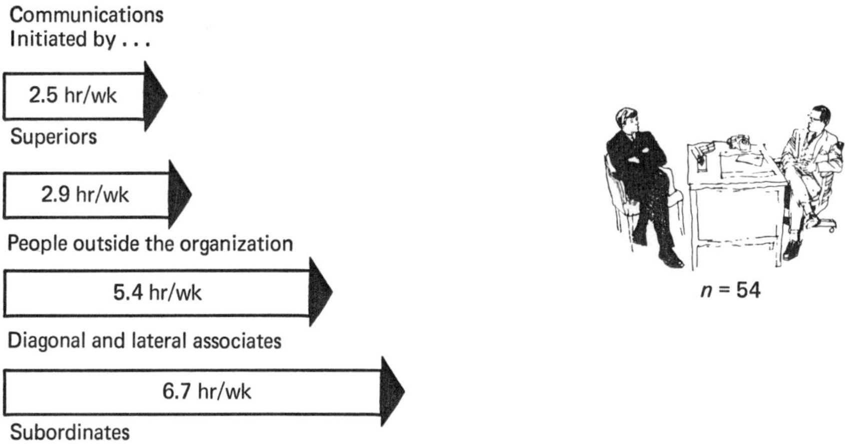

Distribution of Incoming Communications (mean hours per week)

These communications are conducted in a variety of forms: (1) two-person verbal communication whether face-to-face or on the telephone, (2) verbal at group meetings, and (3) written.

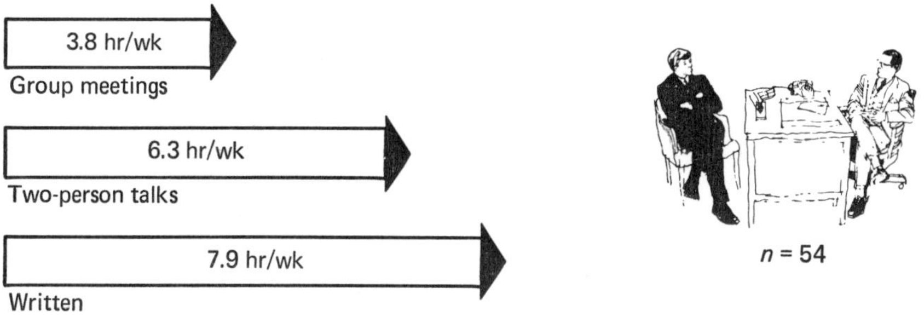

Modes of Incoming Communications (mean hours per week)

Clearly, more total time is spent on verbal communications, but reading written and printed incoming matter is substantial.

Outgoing Communications

Of course managers also initiate communications. One of the most popular images of management shows an executive calling someone on the phone with instructions. The image is exaggerated. Like incoming communications, outgoing ones are not entirely one-way; they include conversations that the manager initiates.

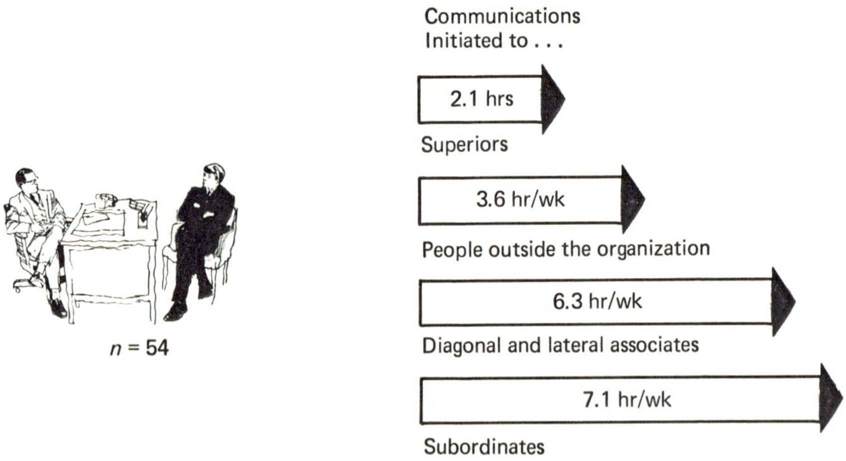

Distribution of Outgoing Communications (mean hours per week)

The typical manager spends slightly more time on communications that he initiates than on those he receives (20.0 vs 17.6 hr/wk). As with incoming communications, more are initiated to subordinates than anyone else. However, they are closely approached by communications initiated to diagonal and lateral associates. Relatively few communications are directed to the manager's direct superiors.

It is striking how the verbal mode predominates outgoing communications in comparison with incoming. Over three-quarters of the time devoted to outgoing communications is verbal compared with about one-half for incoming. The typical manager apparently spends relatively little time initiating written communications.

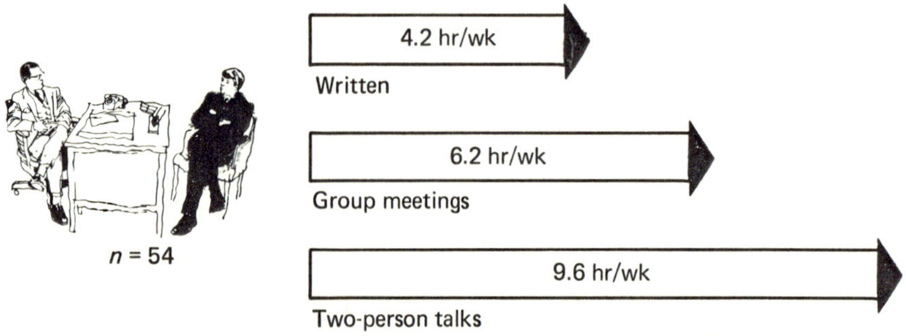

Modes of Outgoing Communications (mean hours per week)

Once again, two-person interactions are the most common form with substantial time devoted to convening and conducting group meetings.

These results corroborate what most managers are well aware of: most of their work is with people—they must talk a lot, and they must have energy to maintain a high-activity level. Nonetheless, the popular image of the manager as an incessant interpersonal activist is a little exaggerated. Substantial time is devoted to quiet reading, analysis, and thinking. As we shall see, however, this time may be too fragmented.

So one stereotype of the manager—that of the verbal communicator initiating and responding to people—is at least partially supported. Another is not, however. Organization charts and principles of management like chains of command, unity of command, and span of control suggest that most communications would be with superiors or subordinates. For the typical manager, however, it appears that almost as much time is spent with lateral and diagonal associates.

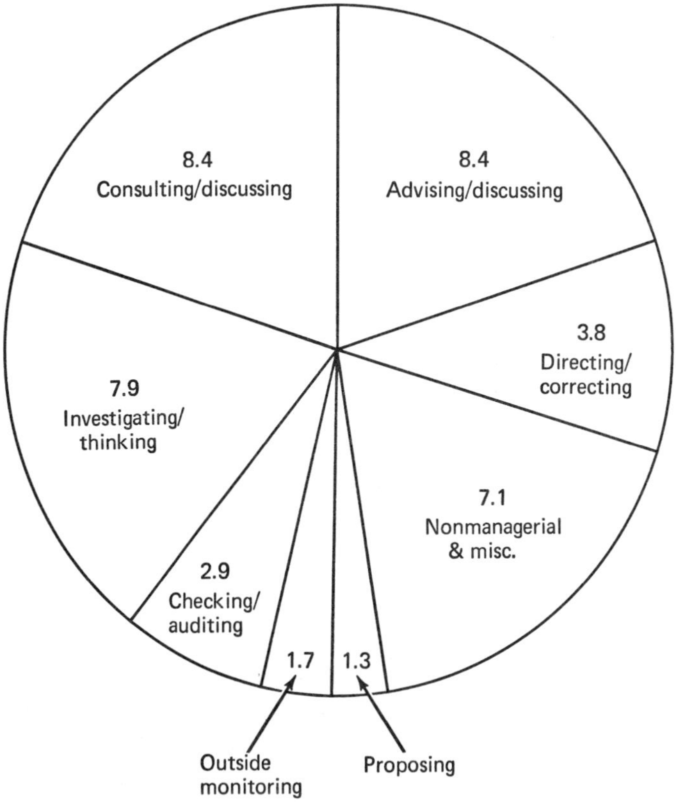

Distribution of Time on Behavioral Elements (mean hours per week [n = 54)

ELEMENTS OF BEHAVIOR

But what is the subject of all this communicating? What is the purpose? Utilizing some common behavioral elements, let us examine the behavior of the sample of managers.

A rather close balance is evident among responding to others with advice and discussion (8.4 hr/wk), initiating consultations and discussions (8.4 hr), and sitting in the office investigating and thinking (7.9 hr). The authoritative functions of checking subordinates' work, directing and correcting occupy a smaller portion of the typical manager's time (6.7 hr). Miscellaneous activities are mainly nonmanagerial; these include production work, technical, clerical work, and selling (7.1 hr). It is a little misleading to consider the typical manager here, however, because most managers do not sell, nor do all spend much time on production/operating work. We shall discuss this more in the next chapter.

PERCEPTIONS OF BEHAVIOR

Obviously most managers are busy people. One result of this activity is that their perceptions of their own behavior are distorted. They tend to see themselves as being busier than they really are, doing more of certain things then they actually do, and especially communicating more with subordinates than the latter perceive.

For example, here is how sixty pairs of managers and subordinates perceive their communications with each other:

Average superior perceives:	4.1	hr/wk communicating with the subordinate
Average subordinate perceives:	3.0	hr/wk communicating with the superior
Actual (estimated from sample of observations):	2.8	hr/wk that the superior and subordinate were communicating with each other

Thus, it appears that managers tend to overestimate how long they spend with subordinates.[7] Other studies indicate that they think they tour and inspect their organizational units more than they actually do.[8] Perhaps they feel so harried that they overestimate everything that they do. Or, perhaps they are vaguely aware that their relations with subordinates are inadequate, so they defensively exaggerate their perceptions. In any case, managers should be a little skeptical as to how they think they allocate their time—and recognize that others probably do not see it as they do.

In summary, research on managerial behavior indicates:

(1) The manager's world tends to be verbal—especially with those communications that he initiates.
(2) Much time is spent in nonhierarchical relations with lateral and diagonal associates, as much as with subordinates, and more than with superiors.
(3) Checking performance, issuing directives, and supervising require less time.
(4) A major portion of time is utilized by giving and getting advice and information to and from many different people.
(5) Managers have distorted perceptions of their own behavior. In particular, they tend to exaggerate how much time they spend communicating with subordinates.

REFERENCES

1. For a fascinating discussion of the use of time and space in communicating, see Edward T. Hall, *The Silent Language* (Doubleday, Garden City, 1959), and his *The Hidden Dimension* (Doubleday, Garden City, 1966); both are available in paperback (Fawcett Premier, 1969).

2. Data on hours worked per week from Bureau of Labor Statistics, *Employment and Earnings,* Vol. 17, No. 11 (May, 1971).

3. The estimates on off-the-job hours for executives is from Sebastian de Grazia, *Of Time, Work and Leisure.* (© 1962 by the Twentieth Century Fund.) He presents the following data on the decreasing work week, p. 441).

Length of Average Work Week in Agriculture and in Nonagricultural Industries, 1850–1960

Year	*All Industries*	*Agriculture*	*Nonagricultural Industries*
1850	69.7	72.0	65.7
1860	67.8	71.0	63.3
1870	65.3	70.0	60.0
1880	63.8	69.0	58.8
1890	61.7	68.0	57.1
1900	60.1	67.0	55.9
1910	54.9	65.0	50.3
1920	49.4	60.0	45.5
1930	45.7	55.0	43.2
1940	43.8	54.6	41.1
1950	39.9	47.2	38.8
1960	38.5	44.0	38.0

Kahn and Wiener in *The Year 2000,* p. 175, give the following prediction of average hours of work in manufacturing: 1965—41.1 hours, 1975—41.5 hours, 1985—38.9 hours, 2000—32 hours.

4. Paul E. Holden, Carlton A. Pederson, Gayton E. Germance, *Top Management* (McGraw-Hill Book Co., New York, 1968), p. 196.

5. See Kahn and Wiener, *op. cit.,* p. 218. Harvard Business School professor Theodore Levitt writes: "Thus it is probable that the demands on his off-the-job time will accelerate more rapidly than the expansion of the time itself. Time that is not consumed by the necessities of study and of keeping up with the new uses of knowledge will rapidly be consumed by commitments modern man makes to other patterns of time usage" ["The New Markets—Think Before You Leap," *Harvard Business Review,* Vol. 47, No. 3 (May/June 1969), p. 60].

6. Most of the data cited from my research is from Ross A. Webber, "Managerial Behavior, Personality, and Organizational Structure," unpublished Ph.D. thesis, Columbia University, New York, 1966. The fifty-four managers studied included a wide variety of staff and line positions ranging from line foremen to vice-presidents of finance, production, and general manager. Thus, there are managers sampled from first-line management upward through five hierarchical levels. I am indebted to Dr. Eliot D. Chapple, Organization Research Laboratory, Noroton, Connecticut, for access to the source material from which the data are derived. Some other surveys and detailed reports of managerial behavior are Leonard R. Sayles, *Managerial Behavior* (McGraw-Hill Book Co., 1965); Robert Dubin and S. Lee Spray, "Executive Behavior and Interaction," *Industrial Relations: A Journal of Economy and Society,* Vol. 3, No. 2 (February 1961); Robert Dubin, "Business Behavior Behaviorally Viewed," in George Strother ed., *Social Science Approaches to Business Behavior* (Irwin-Dorsey, Homewood, Ill., 1962); J. Kelly, "The Study of Executive Behavior by Activity Sampling," *Human Relations,* Vol. 17 (1964); and Jill Horne and T. Lupton, "The Work Activities of Middle Managers," *Journal of Management Studies,* Vol. 1 (1965).

7. See Ross A. Webber, "Perceptions of Interactions Between Superiors and Subordinates," *Human Relations,* Vol. 23, No. 3 (1970), pp. 235–248. The differences in perceptions are siqnificantly different at the .05 level.

8. See T. Burns, "The Direction of Activity and Communications in a Departmental Executive Group," *Human Relations,* Vol. 7, No. 1 (1954), pp. 73–87; and Sune Carlson, *Executive Behavior* (Strombergs, Stockholm, 1951).

4

Present Time: Behavior of Managers in Various Positions

Managers occupy a wide variety of positions, each with its own particular demands and responsibilities. As a result, managerial posts differ as to the behavior required and in the capacity of incumbents to control their own behavior. Unfortunately, much advice on time utilization ignores these differences between positions. We shall examine typical behavior in six different kinds of management positions and discuss how they vary as to discretionary control over time allocation.

TYPES OF MANAGERIAL POSITIONS

Moving generally downward in the organizational hierarchy, the positions to be considered are the following:

(1) *General Executives:* This position includes high executives such as presidents, vice-presidents, and general managers in all functional areas.

(2) *Sales Managers:* These are managers who supervise sales activities—for instance, regional and district sales managers.

(3) *Functional Control Managers:* Functional authority exists when

control over some specific activity is delegated to specialized staff departments. This authority is granted in order to utilize the specialist's expertise and to promote standardization and consistency. Accordingly, these are middle-level staff managers in charge of various functional activities in engineering, finance, industrial relations, accounting, and so on.

(4) *Service Managers:* This position includes lower- to middle-level managers in charge of activities that provide service for primary operating departments. They are not service department managers with automobile dealers; they are managers of departments in industry. A typical situation: the manager has the usual office with a door; he has a phone, sometimes a teletype or a computer printout, even perhaps an old-fashioned pneumatic tube. These expedite receiving communications that are mainly requests or demands for services. A man walks in the door, the phone rings, a message is printed out. All want the service manager to provide something that will allow them to proceed. This something could include repairing, testing, drafting, dispatching, and so on.

(5) *Operating Supervisors:* This position includes lower- to middle-level line managers such as foremen and production superintendents.

(6) *Staff Specialists:* Although not usually called managers because they have few direct subordinates, these people are frequently professional in educational background and are intimately involved in managerial decision-making. They sometimes exercise authority. Most of these specialists work for functional control managers, some directly for general executives, and a few for service managers and operating supervisors.

This is not an inclusive list of all kinds of managerial positions, but it does represent a good cross section of common jobs.[1] Drawing on our research, we shall describe the utilization of time by persons in these differing positions.

COMPARING BEHAVIOR

Six dimensions are important in understanding how positions compare as to their utilization of time. These dimensions are not mutually exclusive, but consider behavior in different ways. To display similarities and contrasts graphically, the positions are located along six continuums.

Time on Vertical Communications

This dimension is the total time spent communicating with superiors and subordinates above or below the manager in the chain of command.

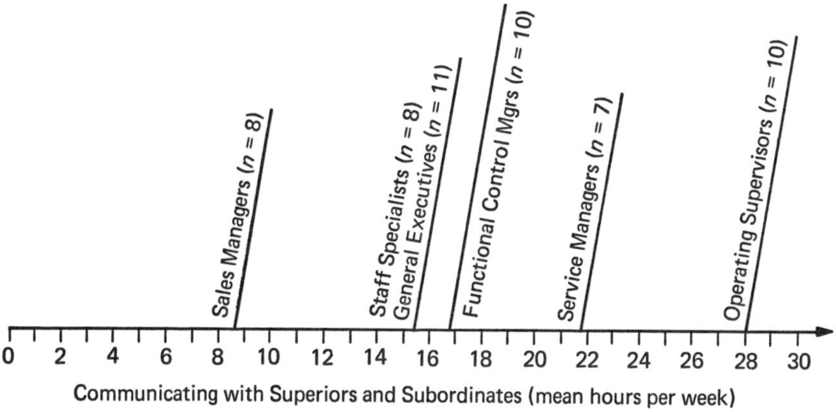

Time on Lateral and Diagonal Communications

This is the total time utilized communicating with lateral and diagonal associates outside the manager's chain of command.

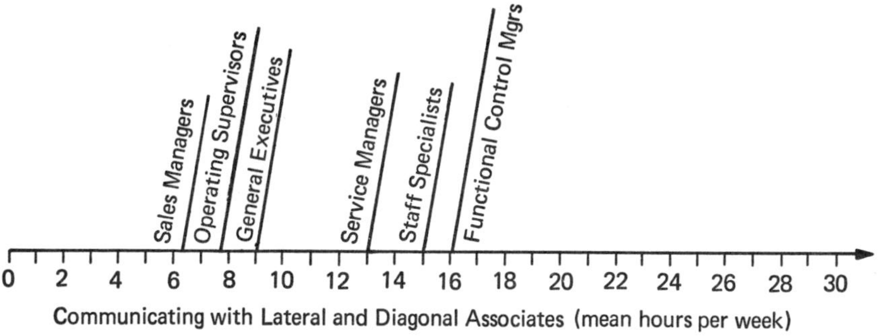

Time on Outside Communications

This is the total time devoted to communicating with people outside the organization—customers, clients, competitors, government officials, professional associations, and so on.

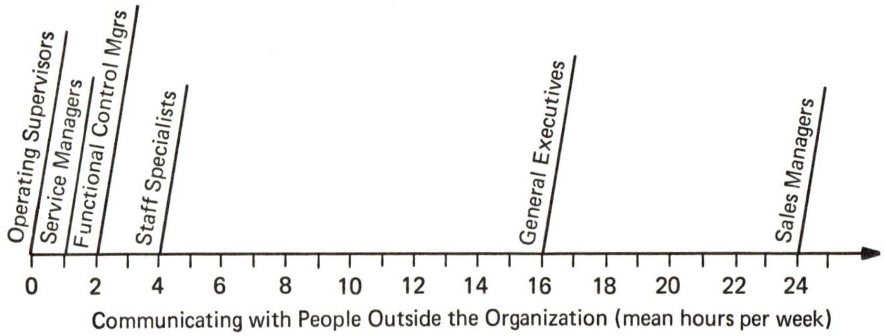

Response Time

This dimension sums up the demands made upon the manager to respond to others. It includes time talking with people who call or come to him and time checking or auditing subordinates' work.

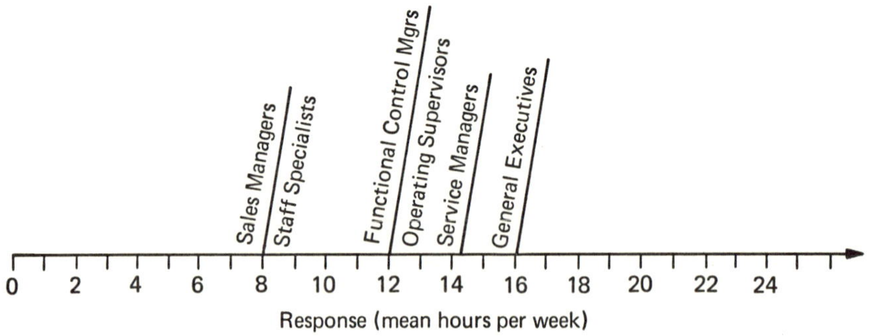

Self-initiated Time

The inverse of response time, this dimension indicates the time devoted to activities and communications that the manager initiates himself. It combines time investigating, thinking, consulting, discussing, directing, proposing, and doing nonmanagerial work—all activities that he initiates.

Directive Time

The authoritarian dimension of management is suggested by time devoted to checking, auditing, and directing the activities of others.

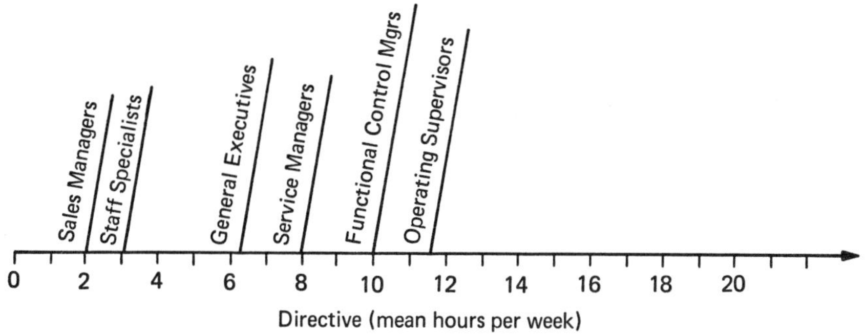

Observations

Several points stand out from these comparisons:

(1) General Executives: Much time, many communications, and a wide variety of people dealt with characterize these high-level executives. In general, they spend more time communicating than any other managers—especially responding to others with advice and discussion. Yet, they tend to be non-directive; they devote relatively little time to checking on or directing subordinates. Finally, they put in more time working than any others.[2,3]

(2) Functional Control Managers and Staff Specialists: These people have the most varied internal communications. They are able to move outside the chain of command to a greater extent. In contrast to operating supervisors, service managers, and sales managers, they have great opportunity to meet various people in the organization and thus to learn its interrelationships and politics. Staff specialists respond little and initiate much; this suggests that their role is less that of adviser than innovator.

(3) Service Managers: These managers respond to associates and supervisors, but initiate to subordinates. They spend much time responding to others who approach them with questions, requests, and demands. Then, they tell their subordinates what to do. In fact, service managers initiate more communications to subordinates than any other position.

(4) Operating Supervisors: These managers are the most exclusively involved within the organization and their own small portion of the chain of command. They communicate least with people outside the organization, most with direct superiors and subordinates, and little with anyone else. This suggests that it is difficult for them to develop a picture of the whole organization. Operating supervisors tend to be the most authoritarian managers; they spend more time checking subordinates' work and directing their activities than people in any other position.[4]

(5) Sales Managers: Relatively little coming in and much going out describes the flow of the sales manager's job. He spends less time talking with people who come to him and more time talking to those he contacts than any other manager. Perhaps this reflects a certain isolation from internal organizational processes—a complaint not infrequently voiced by regional sales managers. They were found to spend the most time communicating with customers' clients (and competitors) and the least time with anyone inside the company.[5]

DISCRETIONARY CONTROL OVER BEHAVIOR

What influences where a manager is located along these behavioral dimensions? Is it individual choice or is he constrained by the nature of his position? This is an old debate. Including Adam Smith in the eighteenth century and Karl Marx in the nineteenth, an impressive collection of social critics have criticized the impact of organizations on individual behavior and personality. Particularly bitter remarks have been offered on how business so limits autonomy that conformity is essential for personal survival.[6] However invalid in relation to real life in business, a widespread image of the corporation has been that of a mechanical apparatus operating impersonally once set in motion by administrative edict; the individual member frequently is regarded as a cog in the machine, what he thinks and does being determined by technological and organizational imperatives.

To many practicing managers, these views on the awful power of the organization seem exaggerated. Undoubtedly they are right. Nonetheless, surveys of college students indicate that this idea that organizational demands dominate individual personality is widely shared. The perceived overrationalization and overcontrol of business is especially resented. So much so that many young people say they are not interested in business careers. Indeed, many apparently want to avoid any career choice that would seem to subject them to powerful organizational demands.[7]

When discussing these matters with managers, we have been impressed by their rejection of the contention that organizational structure dominates the individual. In their zeal to proclaim (and perhaps to exaggerate) their own independence, managers tend to go to the opposite extreme: "Your job is what you make it" is the response. Most seem to feel (or at least to say) that their behavior is pretty much determined by their personal desires and how they want to run their positions. Particularly where a company has promoted a "management by objectives" philosophy, managers see this as support for their view that one can behave as desired as long as goals are achieved.

Thus we have two contrasting views: is the manager's behavior determined by the organization or does it reflect his own personal choice? Common sense suggests that attributing omnipotence over behavior to either structure or personality is too simple. Both influence behavior. We know that demands on people vary with positions. Even on a simple matter like punctuality, there are differences in the extent to which temporal precision is demanded for people in different jobs. In general the higher the status, the less punctuality. Few expect to gain access to the doctor's inner office at the time of our appointment, but how angry we are if a junior subordinate is habitually late for meetings. All positions simply do not allow the same autonomy or discretionary control over one's own behavior.

We can compare the autonomy inherent in managerial positions by examining how widely behavior varies in each kind of post. That is, if all managers in a certain kind of position behave quite similarly, it implies that the job imposes certain behavior. In order to perform and meet the organization's demands, a manager in this position usually behaves in the necessary pattern. In contrast, if managers in a certain position vary widely in behavior, the post would seem to be less structured, the demands less clearcut, so that greater discretionary control over one's behavior is possible.

Analysis indicates significant differences in behavioral variability of managers in the different kinds of positions. They are arranged as follows:

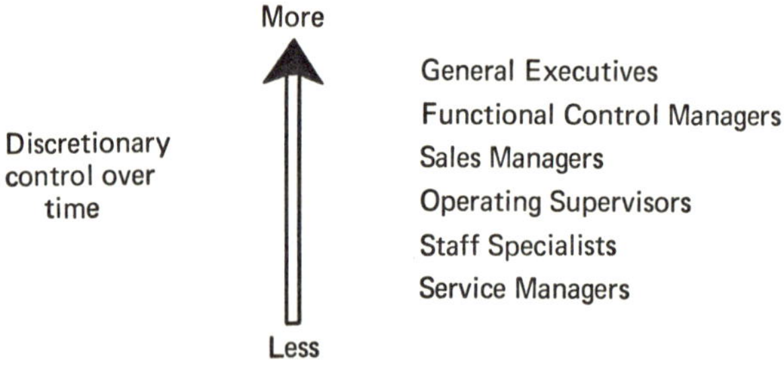

For example, service managers occupy the most programmed managerial position—they demonstrate the least variability in behavior. The job's requirements are so explicit and demanding that relatively little discretion is left to the incumbent. Many requests, directives, and demands come to him. To most of these communications, the service manager responds by categorizing the request and issuing it to a subordinate. This manager's job is primarily responding to others' demands, so the organizationally determined imperatives of his position are great.

In contrast, high-level general executives have great discretion in controlling their own behavior. Behavior varies more widely among these managers than for any other position.[8] Demands are ambiguous, so incumbents vary widely in behavior. Explicit short-range demands are relatively rare—for example, these men apparently make fewer decisions (although more important and more difficult) than service managers or operating supervisors.

AUTONOMY AND LEADERSHIP STYLE

Please note that we do not imply that either supportive or authoritarian leadership is better. Supportive style is indicated by more time responding to subordinates with advice and discussion, less time checking on their performance and issuing directives. Authoritarian leadership is associated with more directing and checking, less responding. Contrasting styles reflect positional demands and the autonomy allowed.[9] Thus, general executives have the most discretionary control over behavior and they tend to be supportive in encouraging others to contact them while refraining from directing. In contrast, operating supervisors and service managers have less discretion because job demands are explicit and constant; they are likely to be authoritarian. Thus, highly structured managerial positions with many explicit demands tend to manifest authoritarian leadership. The key variable is time: the time span required to obtain feedback on performance and the time span between performance reviews. The shorter these time spans, the more structured the position and the less discretionary behavior. Thus, the time orientation of managers tends to vary with the average time to get definitive performance information.[10]

Most men desire autonomy because it recognizes their maturity. This desire underlies the call for collaborative and participative management and the effort to expand discretionary behavior. Some suggest that management should go further in fostering autonomy in modern organizations. Even matters like working hours and vacations may offer possibilities for increasing autonomy. "Everyone over $150 per week should set his own hours" proclaims former Avis president Robert Townsend.[11] He maintains that such people should have the freedom to determine their own hours and schedules because they have different working patterns.

It is difficult to judge whether this is a feasible proposal, but some anthropologists do suggest that the most natural work pattern is intensive, prolonged activity followed by extensive rest. The regular, limited schedule that characterizes business today was developed mainly to suit organizational and technological requirements, not to satisfy human needs. Men still differ as expenders of energy: some want to work steadily at a predictable, fairly moderate level; others feel frustrated if they cannot alternate between furious expenditure of energy and periods of idling. It is a weakness of modern organizations that they offer comparatively few opportunities at other than high levels for the second type of temperament. Men of this kind may far outnumber the opportunities that modern life can offer them.

Nonetheless, such flexibility is difficult to implement. Unless employees work quite independently, as in some research laboratories, individual choice presents problems (even in company labs, nonprofessionals are sometimes offended by the researchers' freedom to choose their own hours). Where technology requires predictable presence and where people are dependent upon each other for integrating their activities, such autonomy seems impractical. Still, we need to explore ways of increasing individual autonomy within organizations.

REFERENCES

1. The general executives studied included such positions as Vice-president and General Manager, Vice-president of Finance, Assistant Vice-president of Western Sales, Vice-president and Comptroller, Chief Engineer, among others. The functional control managers included Planning and Control Manager, Auditor of Revenues, Plant Laboratories Manager, Director of Industrial Engineering, and others. Service managers included Electrical Laboratory Head, Methods and Cost Reduction Head, Testing Superintendent, Field Department Head, and others. Staff specialists included Real Estate Specialist, Senior Transportation Engineer, Chemical Engineer, Shipping Specialist, Designer, Assistant to Director of Development, and so on.
2. The classic, early study of high executives was Sune Carlson, *Executive Behavior.* A study of young presidents is S. Klaw, "The Entrepreneurial Ego," *Fortune,* Vol. 54 (August 1956), pp. 100 ff. My findings are not dissimilar to those of Carlson's pioneering study. He writes: "They clearly showed, for instance, that for the chief executive the main problems were those of getting information and advising and explaining. In their opinion, most of the directors studied did not take part in so very many decisions and it was seldom that they gave orders," and ". . . the main problem for the executive in dealing with questions brought up is to keep himself informed." (p. 49)
3. Sebastian de Grazia comments on the fact that higher executives seem to work longer than anyone else: "The executive may not be able to tell work from play; nonetheless, he has little free time, perhaps the least of all categories of persons we have examined. The class that is able to wrest the highest monetary rewards out of the economic system and close to the highest rewards in prestige cannot or doesn't want to have the most free time. When in the 19th century, England led the world in trade and finance, London executives took 4-day weekends. If executives are so powerful a force in America, as they indubitably are, why don't they get more of that free time which everybody else, it seems, holds to be so precious? Strange." Perhaps de Grazia overlooks the satisfactions that the executive derives from his work. (*Of Time, Work and Leisure,* p. 138; © 1962 by The Twentieth Century Fund.)
4. For other studies of foremen, see: F. J. Jasinski, "Foreman Relations Outside the Work Group," *Personnel,* Vol. 33 (September 1956), pp. 130–136 and Robert H. Guest, "Of Time and the Foreman," *Personnel,* Vol. 32 (May 1956), pp. 478–486.
5. On Sales Managers, Robert Townsend, the former president of Avis, writes: "Most good salesmen thrive in the field, wither at headquarters.

'There I was alone,' a true salesman once said, 'with nothing but my golden voice.' " (*Up the Organization,* Alfred A. Knopf, New York, 1970), p. 166.

6. For more on the impact of organization on the individual, see: William H. Whyte, *The Organization Man* (Simon & Schuster, New York, 1956), Alan Harrington, *Life in the Crystal Palace* (Alfred A. Knopf, New York, 1959), C. Wright Mills, *White Collar* (Oxford, New York, 1951), Chris Argyris, *Personality and Organization* (Harper & Bros., New York, 1957), Lyman M. Porter, "Where Is the Organization Man," *Harvard Business Review,* Vol. 41, No. 6 (November/December 1963), pp. 53–61, and Leonard R. Sayles, *Individualism and Big Business* (McGraw-Hill Book Co., New York, 1963).

7. See the *Newsweek* survey mentioned in references to Chapter 1.

8. On autonomy, Sune Carlson writes about his executives: "The content of their working day is determined only to a small extent by themselves, and it is difficult to change it without making considerable alterations in the organizational structure of which they are part. Before we made the study, I always thought of a chief executive as the conductor of an orchestra, standing aloof on his platform. Now I am in some respects inclined to see him as the puppet in a puppet show with hundreds of people pulling the strings and forcing him to act in one way or another." (*Executive Behavior,* Strombergs, Stockholm, 1951, p. 52) My research suggests in fact that high executives have greater freedom to shape their behavior than any other managers.

Former Assistant Secretary of State Charles Frankel has written of his first awareness that high office limited one's freedom as well as expanding one's power: "The taste of power, or whatever it was that I tasted the first day, went to my head too, but not quite as I had been warned it would. I had come into the office with projects and plans, and I was caught in an irrestible movement of paper, meetings, ceremonies, crises, trivialities. There were uncleared paragraphs and cleared ones, and people waiting for me to tell them what my plans were, and people doing things that had nothing to do with my plans. I had moved into the middle of a flow of business that I hadn't started and wouldn't be able to stop. There were people in place to handle this flow, and established machinery in operation to help me deal with it. The entire system was at my disposal. In a word, I had power. And power had me." (*High on Foggy Bottom,* p. 4)

9. There is a large literature on situational or contingency leadership. Perhaps the best known is Fred E. Fiedler, *A Theory of Leadership Effectiveness* (McGraw-Hill Book Co., New York, 1967); also Robert Tannenbaum and Warren H. Schmidt, "How to Choose a Leadership Pattern," *Harvard Business Review,* Vol. 36, No. 2 (March/April 1958), pp. 95–101; and John J. Moorse and Jay W. Lorsch, "Beyond Theory Y," *Harvard Business Review,* Vol. 48, No. 3 (May/June 1970), pp. 61–68. A fairly complete review of recent research may be found in *Administrative Science Quarterly,* Vol. 16, No. 1 (March 1971).

10. The impact of time spans on orientations is examined by Paul R. Lawrence and Jay W. Lorsch, "Differentiation and Integration in Complex Organizations," *Administrative Science Quarterly,* Vol. 12, No. 1 (June 1967).

11. Townsend, *op. cit.,* p. 131.

5 Why Time Is Wasted

Literature on time management tends to portray the manager as an unarmed rotund merchant passing through a thicket "filled with surreptitious time robbers." Surveys of managers indicate the following factors as common thieves lurking in the threatening forest:[1]

- meetings — excessive length because people are unprepared; high-level meetings on minor problems
- visitors and unanticipated interruptions; talking with lower subordinates who have bypassed the chain of command and who want to talk rather than just give results of their work
- long letters
- failure to delegate
- routine detail
- poor organization
- procrastination
- unnecessary correspondence —especially unread reports
- telephone interruptions—especially the priority given to incoming phone calls over people in the office
- keeping up with former position after being promoted
- misplaced items
- commuting
- waiting for people
- mediocre personnel
- correspondence delays
- coffee breaks

- trivial matters (cited are communicating with stockholders, entertaining head office visitors, internal public relations, and office politics)

Such items are not unfamiliar to most managers. They recognize some of their personal villains there. Nonetheless, most of the items are self-centered and superficial; they represent symptoms, not causes, and imply that anything that interferes with the way managers want to spend their time is wasteful. Managers must deal with themselves, not the bandits of Sherwood Forest. Besides, as Chapter 4 indicated, not all managers can or should behave similarly; different positions require different behavior. What is waste for one manager is not necessarily waste for another.

Much of the trivia that managers complain about as time-wasting really consume a trivial amount of time. The big losses stem from more fundamental problems: too much job, too much time, too much secrecy and too much fear.

TOO MUCH JOB

Some positions, of course, just contain too much. The incumbent can be overwhelmed by the variety, quantity, and urgency of demands. In order to deal with this stress, a manager may develop protective mechanisms that jeopardize job performance.[2] Even the best of us can stand only so much activity and pressure on the job. Under excessive demands, a manager departs from his characteristic behavior pattern: he may lower his activity level by reducing responsiveness to incoming stimuli; or he may attempt to drown out external demands with his own hyperactive behavior; or he may withdraw and isolate himself. All of these actions under stress result in changes in volume, direction, and content of communications.

Thus stress sometimes is accompanied by stepping up communications The disturbed manager may increase initiation to subordinates and associates—usually resulting in more, but shorter, interactions. "Whenever the Operations Officer became upset," comments a former naval officer, "he would soar to a high activity level. Talking and moving about a great deal, he would go from one compartment to another, giving orders, shuffling papers, and chewing men out."

For other managers, stress may be indicated by progressive withdrawal. Avoidance of communication with a specific individual is described by the same officer: "The Executive Officer's job was to check on my performance. I never liked having to respond to him . . . I soon decided that to control my temper I should avoid him as much as possible. Inspecting the ammunition magazines in the ship's holds became my excuse." Similarly, under stress a food store manager cut down on his informal relations with his men; he required them to contact him "through the proper channels." And with production going badly, an electronics operating supervisor isolated himself to repair defective components himself and to avoid complaints and criticism. This ineffective supervisor spent half as much time communicating as a more successful supervisor in a parallel post.

With such disturbed communications, managers tend to overemphasize certain relationships and slight others. A common result: overemphasis on vertical relations. But, every job needs contacts outside the chain of command, for a manager who confines his communications to immediate supervisors and subordinates is likely to limit his attention to short-run problems. He will ignore more basic technological and structural issues and concentrate on pushing his people harder (or getting rid of "troublemakers"). This condition is illustrated by the ineffective electronics supervisor cited above who had virtually no contacts with staff specialists or other production groups. Similarly, low-rated supervisors in an automobile assembly plant constantly initiated to subordinates and resented any interaction with staff personnel because of the pressure of short-range problems.[3]

Such distorted communication flows are also often accompanied by changes in what is said and how. Common are more commands, fewer questions, and less informal discussion. When conditions were chaotic in the automobile plant mentioned above, one-way commands constituted almost all communications. No one even had time to ask what was wrong. Reducing social conversation would seem to save time and make sense, but management may also lose information that often is gained spontaneously from such small talk.

Excessive demands are caused by failure to draw adequate limits on jobs.[4] For high-level executives these demands frequently come from outside the organization because they tend to personally handle many matters originating outside the firm. Except for customer orders, intentional delegation is less apparent in this area; they seem to accept outside issues without enough differentiation between the important and the trivial. At present, many executives are extremely busy responding to external pressures on pollution control and product quality. These are very important subjects, but not every relevant item warrants high-level attention. The problem is that top management has been unable to convert these pressures into operational objectives for lower personnel. In a public utility company, the executives expend much time correcting projects so that they reflect greater awareness of environmental impact. Unfortunately, this concern has not yet been effectively converted into criteria by which the design engineers can guide and evaluate their own performance. Hence, they are still emphasizing mainly economic and engineering factors. Inadequate translation of organizational objectives down to unit goals is probably the major cause of too much job remaining at the top.

Excessive demands on middle management sometimes stem from higher management's failure to define the former's jobs and to state clearly what is expected. Middle management may not know where their responsibility stops and someone else's begins. Managers should be told where their responsibility ends because in a competitive business, where rivalry is acute, ambitious and conscientious men will cover every possibility. They may overwork themselves unnecessarily, intrude into others' areas, and work endless hours in protective effort.[5]

So excessive demands lead to job stress, distort managerial behavior, and waste time. The effect is evident in the problems cited by one study of high-level executives:[6]

(1) Inadequate time for reading—As a result, executives tend to be narrow in their interests and overly concentrated on technical and economic reading.

(2) Inadequate time for internal inspection and visiting—Executives feel that this is important and regret not doing more, but they still overestimate how much they actually do.

(3) The capriciousness of their time—They tend to have only short intervals of uninterrupted time insufficient for concentrated thought and contemplation (the average span of time alone was 14 minutes—and if telephone interruptions are counted, the average period without interruption was only 8 minutes). Five or ten minutes alone are helpful only for work on small, immediate problems. They are unsuitable for long-range matters and even too short as rest intervals. Many executives did not believe that they spent up to an hour or hour and a half "alone" during the day. All they knew was that they scarcely had time to start a new task or light a cigarette before they were interrupted by a visitor or telephone ring.

TOO MUCH TIME

How strange that one of the biggest wastes of time results from having too much of it! This is a conclusion reached from talking with many middle-level managers—especially in staff positions. Many managers have imposing titles, fancy offices, apparent status, but inadequate job demands because of overstaffing, little delegation, or depressed production and sales. Unless they are uncharacteristically passive, such managers want to act; a desire that results in time devoted to unnecessary activity. An underchallenged manager may compensate in three ways:[7]

(1) He may increase his activity by creating unnecessary problems which he can then solve. Conflict lies close to the surface in every human group. For example, some psychologists suggest that the high divorce rate in California partially reflects the beautiful weather and good living. Without the Northeasterners' external problems to worry about, real but minor family stress is exaggerated. Oversensitive human relations and personnel difficulties are common products of overstaffing and excess time.

(2) He may increase his activity by overreacting to a normal stimulus. Most familiar are such habits as continual grooming, scratching, nail-biting, but also included are various occupational devices that serve as time killers: elaborate and long luncheons, inappropriate attention to the format and mechanics of reports, unnecessary reorganizations, and so on. The most pervasive symptom of excess time and subsequent overreaction, however, is excessive meetings. Meetings of course are often helpful, but they tend to proliferate and lengthen when too many people are concerned about too few problems.

(3) He may increase his activity by normal or exaggerated responses to subnormal stimuli. We are all familiar with people who interpret everything in terms of what others think about them. A friendly smile from a secretary becomes an invitation to date for some Walter Mittys—especially if they are bored. A clerk arriving five minutes late may become the occasion for a crackdown on tardiness by public humiliation of the poor girl—who might have been held up by an inoperable door on the Eighth Avenue subway.

C. Northcoate Parkinson lent his name to these phenomena in the now famous Parkinson's law—that a job expands to fill the time available.[8] Actually, this is not exactly true. If it were and personnel perceived their work as legitimate, excess time would not have the deleterious effect on morale that it usually has. A few years ago, Sandia Corporation, primary designer and tester of nuclear weapons for the Department of Defense, was faced with a large surplus of labor. After the cessation of atmospheric testing, perhaps one-half of its highly skilled work force of 8,000 scientists, engineers, technicians, and craftsmen were not needed. They were not released, however, because the government was willing to carry them in order to maintain the organization's capacity intact. Nonetheless, the professional and managerial personnel could not fool themselves by expanding their limited work to fill all time. The empty stretches of time were very tough to handle. Achievement-oriented managers became frustrated and status-conscious people became nervous; all became uneasy about their worth as professional men. Morale deteriorated, *as did performance even on the necessary and important projects.* Man has a great capacity for self-delusion, but it is not unlimited. Most of us cannot completely fool ourselves that limited work can be expanded legitimately to fill excess time.

Some disturbing statements were recently made by the former commanding officer of the United States Navy destroyer that was reportedly attacked in the Gulf of Tonkin in 1964.[9] This attack led to the famous resolution by the Senate that was used to explain our military activity in Vietnam. In a newspaper interview, Commander Ogier of the "Maddox" told about his growing doubt that his ship had been the object of a sustained torpedo attack from North Vietnamese naval units. Two of the reports that torpedoes had been launched were perhaps authentic, but nineteen subsequent reports may have "resulted from our putting our rudder over, you see, and this was reflecting the sounds of our own screws up toward our sonar." One wonders how much of what a manager hears is really the sound of his organization's screws reflecting off his own rudder and coming up through its own highly selective sonar.

Perhaps the data on working hours in Chapter 4 suggest a reason why some managers have inadequate stimulation and excessive time. It appears that the higher a manager climbs up the hierarchy, the more time he puts in. This is a remarkable reversal of history. The most pervasive definition of success worldwide has been freedom from work. Yet, here and now the president works longer than the high executive, the executive more than the middle manager, and the manager longer than operating supervisors. We are unaware if this pattern aids or hinders organizational performance, but it does suggest that many businesses are highly centralized—for all the pious talk about decentralization as an ideal. People higher in the organizations do more of the work and receive more of the financial and psychological satisfactions.[10]

The middle manager may have more leisure, but he is not necessarily happy about it. If he has a time-haunted need for achievement, he does not want the free time nor does he know how to use it. He feels guilty about having empty time on the job. More important, if he is ambitious he wants to behave like his boss, but not too differently from his peers with whom he is competing. So he must fill up his time on the job; and this can be time-consuming and boring. Consider the following comments by some managers in the home office of an insurance company:

- "The officers are the first here in the morning and the last to leave at night; they are always here Saturdays and many Sundays."
- "They set the pace and, at least implicitly, it is the pace we must accept and follow."
- "If you want to get ahead this is the pattern you must accept . . . Contribution tends to be judged in terms of time spent in the office, not things accomplished."
- "If you want to get ahead, you come in on Saturdays regardless of whether it is necessary or not. The cafeteria and offices are sometimes filled with people who just feel they can't afford not to come in on Saturday."

As one newly appointed division manager put it, "Holy cow, what's going on? I've looked out of the window so many times a day, I'm about to go nuts. I've gotta do something. What can I do?" Of course, not all or even most managers are characterized by too little stimulation, not enough to do, and too much time. Nonetheless, this is a much more common situation than most companies care to admit.

TOO MUCH SECRECY

Seeking and exchanging information consumes much time. As Chapter 3 indicated, such behavior is central to all management; it is obviously not inherently wasteful of time. Yet inability to gain information because of restricted access and monopolized control is all too frequently a cause of wasted time. Restricted access sometimes reflects a valid need for security, but just as frequently it stems from insecurity and fear.[11] For some young staff professionals, an early stressful experience is finding that fellow employees and managers of the same corporation will not divulge data essential to his assigned project. Panic stricken, the new staff man may resort to stolen peeks at forbidden files while the line manager is out of the office, fearful that he will be found out and his tenuous relationship with the manager completely ruptured. Elaborate espionage schemes are sometimes concocted, not to make anyone look bad, but just to perform his job.

Excessive withholding of internal information is a clue to the character and atmosphere of a company. In ideal organizations, authority and ability happily correspond in each hierarchical position. When ability does not match assigned authority, however, monopolizing information is a way of handling one's insecurity. To possess preferential information is to possess power to make others dependent on oneself. It can limit their autonomy and scope restricting them to predictable and non-threatening actions.

Information is often withheld to forestall questions about an arbitrary decision that has no rational defense. Thus lower level wages are made public through announcement of union contracts and high executive incomes published in stockholder reports, while the veil is drawn closely around middle management salaries. It would appear that the more subjective the reasons for a decision, the greater the inclination to hide both the decision and the information on which it was based.

Much organizational energy can be expended in the struggle for information. Time is consumed by attempts to hide or obtain it. More damaging, the psychological climate is undermined by a nagging belief that others are withholding information that one needs or that some closely guarded secret will leak. Making all information available is obviously impossible; yet, the goal should be pursued. A general characteristic of the time-conscious organization should be to give much wider and more ready access to information than is now usual.

TOO MUCH FEAR

More than too few or too many demands, fear is the greatest reason for wasted time. The current cry of some youth and professors that our organizations are permeated with fear is probably true. Unfortunately, this condition may have less to do with the nature of these organizations than with human nature. Even in this day of chromium, air-conditioned affluence, a man with a job is exposed. And managers may be even more exposed, for contemporary business seems to have reversed the historical balance of organizational fear. In the early industrial revolution and after, fear was management's primary device for motivating blue collar workers—fear of income loss, dismissal, starvation. Now, however, union protection, government legislation, unemployment compensation, welfare, and most important, relatively full employment can rightfully and thankfully blunt the impact of job insecurity on workers (perhaps to be replaced by greater fear in other aspects of life). Paradoxically, the middle manager may be even more susceptible to fear than ever. He has more invested in the job because more depends on it—his psychological satisfactions, social esteem, his living standards with their high proportion of fixed expenses. His greater specialization means a higher income with his present employer, but may also lessen his attractiveness to others (especially if he is over forty).[12] Finally, managers may also be more ambitious to advance; this means competition, and that fosters fear.

How three fears affect the utilization of time will be discussed: fear of losing status, fear of ambiguity, and fear of losing control.

Fear of Losing Status

Occidentals frequently attribute to Orientals an overconcern with losing face. Perhaps they protest too much. Protection of present prestige and promotion of future status are major time consumers in most organizations. Managers expend time promoting themselves; some devote time to undercutting others; perhaps even more time is used in trying *not* to threaten others. Such behavior results from the system's defensiveness and ironically serves to reinforce the same defensiveness. Some call it "windmilling." Others simply state that they spend "a hell of a lot of time doing nothing that moves things forward," or, "Spending time keeping someone happy, but not getting the job done." Estimates of how much time is spent windmilling range from 20 to 50 percent of management's time.[13]

Despite complaints that windmilling lacks purpose, it is intended to be self-protective. The examples are many:

- *Meetings:* Managers frequently gripe about endless meetings. Part of the problem is the investment that the convener and attendee make in a meeting. Almost everyone wants to avoid offending the person who calls the meeting. Most want to perform well in a competitive setting. Consequently, they want to ask at least one question demonstrating their interest and make a statement displaying their knowledge—and these are sometimes uttered even when not pertinent to the meeting's topic.

- *Too many managers:* In many staff departments superfluous managerial levels have been created in order to provide promotions for professionals. Not having much to do (and since not being busy suggests low status), the new managers begin to communicate and supervise, thus effecting a new time drain for everyone.
- *Male-female relations:* A persistent theme among female managers is the care they must exercise not to offend the ego and status of their male superiors, associates, and subordinates. Tact and time are consumed by the indirection required to promote new ideas. Female managers develop subtle techniques for protecting and manipulating their male colleagues into thinking something is their idea.
- *Control systems:* When objectives are not achieved too many managers emphasize finding *the* guilty person who made *the* mistake. Such personalization and simplification encourages time-wasting deception and confusion. When something goes wrong, managers tend to expect the responsible supervisor to know why and what he intends to do just as soon as he is asked. How many managers have had their boss ask, "Why is such and such wrong? What are you doing about it?" Sometimes these questions are thrown out even though the questioning superior has received the control report before his subordinate! Now if a person is expected to know all the reasons and have ready replies, he is likely to conceal problems in order to play for time. Such secretness may isolate him from advice and allow the problem to worsen.

Consider the following excerpt from a T-group session in one company:[14]

> *Norm:* I have a clue. I have a clue. When someone says we're in trouble, our first question is, "What have you done about it?" Now if you've had to have done something, then people aren't we stifle facts because people won't report something they hide information, sometimes until it's too late.
>
> *Lou:* Also, we ask *who* made the mistake—who did it—fire the guy.

Dave: This is important for me. Good point. We're saying that we stifle facts because people won't report something they haven't figured how to lick. How can we create a climate so that he can tell us when the problem occurs?

Fear of Ambiguity

Jews, blacks, and other unfortunate minorities have long been convenient scapegoats for those in fear of society's complexities. The world appears simpler if one assigns people to distinct categories and then ignores the individual. For example, if you don't bother to learn how clean, industrious, and patriotic is *this* particular long-haired bearded student, your hippie stereotype is not disturbed and you can maintain the illusion that you understand current university turmoil. Research on the authoritarian personality indicates that certain people categorize in this way.[15] Psychologists have defined and measured a characteristic termed tolerance for ambiguity. One test consists of flashing a series of small light dots on a screen. Initially everyone agrees that the dots form a dog. Then the light configuration can be slowly altered so that the figure changes. People with low tolerance for ambiguity tend to lag in perceiving and recognizing a new shape; they hold on to the perception of the dog. In contrast, subjects with higher tolerance for ambiguity can relinquish the familiar shape, tolerate a period during which they are ignorant of the shape, and recognize sooner the emerging form of a cat.

To some degree, however, everyone has limited tolerance for ambiguity; we tend to prefer the known (especially if it is acceptable) to the unknown. We prefer the act that has measurable results to the intangible or unmeasurable. This leads managers to concentrate on those details about which they complain so loudly.

Every oraganization has an inexhaustible supply of detailed issues calling for attention—*now!* "I waste my time on trivia," managers complain. "If I could get rid of such nonsense, I would spend my time on important planning and evaluating" is their plea. Perhaps they mean this, but many really want the demand for detailed activity. It is not that managers are forced to expend all their time on trivialities; the opposite may be true. The real challenges of managing are so difficult and anxiety-provoking that managers allow their days to be filled with detail and trivia. At least, when they issue a directive modifying travel allowances, they "accomplish" something tangible, real, and comforting. For some executives, the real challenges of managing are so threatening that they keep busy all week on this sort of thing. This protects them from having any time to contemplate what they are *not* doing.[16]

This attention to familiar detail has special impact when a manager is promoted. Sometimes he continues to hold on to his former position by close supervision, partly because he lacks trust in his replacement, but mainly because he seeks the comfort of the familiar. Managers tend to concentrate where they know they are proficient. Not wrong in itself, such behavior emphasizes the present at the expense of the future and the old to the detriment of the new.

Fear of Losing Control

No place is this fear of ambiguity and its effect on time utilization more evident than in controlling subordinates. Collaborative or participative management includes a period when the superior does not know what his subordinate is doing—or at least he knows only what the subordinate chooses to reveal. For a time after the desired objectives have been agreed upon, collaborative leadership means that the subordinate is free to explore his own means. For example, after superior and subordinate agree that the latter needs a cost accounting and control system in his department, and the superior sets limits upon what can be spent in setting it up, he must give the subordinate freedom even though the superior may be an expert on such plans. The superior should live with ambiguity and allow his subordinate to suggest his own proposal. Why wait for this? Because the subordinate will probably make his plan work more effectively *even if it is intrinsically inferior to an expert's just because it is his.* The superior should create an environment where he is informed voluntarily, but he should refrain from probing for a reasonable time. The superior may be anxious—especially if his boss is exerting pressure because his authoritarian concept of management requires a manager to know exactly what his subordinates are doing at all times. A manager needs courage to demonstrate collaborative leadership when his own superior is authoritarian.

Under threat from superiors and fear of ambiguity, many managers overcontrol. Consider the following example from a training session with some state employees: Three hierarchical levels were present: the state director of correction, the deputy director in charge of prisons, and the various prison wardens. The deputy reiterated the old organization principle that responsibility cannot be delegated, that he would hold each warden personally responsible for anything his guards did wrong (from supplying a car for an escaping inmate to smuggling in forbidden copies of *Playboy*). The wardens were well aware what this implied for their preceding discussion of delegation and decentralization. It was rendered irrelevant. The deputy director's dictum would require extensive monitoring of guards because the most effective way to control closely is to interject yourself between subordinates so that most communications pass through you. This may or may not be effective prison management (although the wardens felt that it would have relatively little effect on whether or not a given guard would collaborate), but it certainly consumes time and sabotages decentralization.

The desire for control and certainty leads to some strange inversions of activities. In one small printing concern, customers coming in the front door were the life of the business. There were neither outside nor inside salesmen, no mail orders and few phone orders. The person who opened the front door and jingled a little bell *was the* business. Yet, surprisingly the proprietor was clearly annoyed by their arrival. Customers were perceived as interruptions to his paperwork—completing job tickets, estimating, calculating, and so on. As much as managers complain about paperwork, it is tangible and controllable. They can start and stop as they please, take a drink of water when thirsty, speed up or slow down. In contrast, the customer is not under control; he initiates the conversation and tries to manipulate the manager to do what he wants. Of course to serve him is the function of the business, but instinctively a manager may prefer the known, the concrete, and the self-controlled.

Fear of ambiguity and losing control also contributes to a crisis syndrome. Of course legitimate crises imposed by the environment or technology can lead to effective performance and high morale (recall how many people enjoyed World War II—when they were not being shot at). For example, under pressure from President Kennedy to produce equipment to detect Soviet nuclear tests in the early 1960s, a crisis atmosphere permeated Sandia Corporation. In a short time, substantial technological innovations were produced which enabled the United States to agree to the atmospheric test ban treaty without inspection. This crisis was dramatic and clearly necessary; it gave direction and meaning to an organization.

In contrast, many crises are artificial—like the "hurry up and wait" of military fame. Such fraudulent conditions seem to grow from the hierarchical structure itself. A high-level executive feels that a completion date is necessary to keep the organization on its toes; he assigns one that he thinks reasonable. Each subordinate manager wants to impress his superior so he advances the date a little. Thus time pressure increases as the deadline moves lower in the hierarchy. An arbitrary crisis is generated that no one understands but which is a customary state in many organizations.

Fear of losing status, of not being in control, and of being adrift in ambiguity all contribute to the long hours of some managers. So, an oil company president complains, "I don't know why, but everybody around me works too hard . . . my executive group all work too much. I think that's a bit characteristic of business executives in the United States today. If anything, they spend too many hours at work."[17]

A sociologist offers his explanation:

> If the executive tends to picture himself as being constantly at work, it may be in part because he feels under some pressure to give a tangible proof of his contribution. So much of what he does is not capable of being readily evaluated; it does not lend itself to being seen and immediately appraised by his associates. Hours of work (some of it unnecessary and wasteful) become one vivid sign—an outward discernible mark—of the extent of his labors.[18]

REFERENCES

1. A survey of time wasters is reported in Ted W. Engstrom and Alec Mackenzie, *Managing Your Time* (Zondervan Publishing House, Grand Rapids, 1967), p. 215. Another survey is reported in "Time at the Top," *Wall Street Journal,* August 13, 1968. The most extensive survey (almost 3000 managers) is reported in Leo B. Moore, "Managerial Time," *Industrial Management Review,* Vol. 9, No. 3 (Spring 1968), pp. 77–85. All the surveys report very similar results.
2. See Ross A. Webber, "The Roots of Organizational Stress," *Personnel,* Vol. 43, No. 5 (September/October 1966). For extended discussion of stress and time pressure in the executive's job, see Theodore Irwin (ed.), *What the Executive Should Know about Tensions* (American Research Council, Larchmont, 1966). The final advice offered on "how to beat tensions" is: (1) On the job, take one thing at a time. Under

tension, your work load may seem unbearable and you feel hopelessly trapped. Take a few of the most urgent tasks, first picking those you think are easiest to handle, and pitch into them one at a time. As you complete each item, your tension will ease up without your being aware of it. This device will help pull you out of your breakneck pace—that depressing feeling of being overwhelmed. (2) In the privacy of your office, take a breather. Stand up, inhale deeply and deliberately, and let your breath out slowly. (3) Slump in your seat from time to time. (4) S-t-r-e-t-c-h hard. Roll your shoulders, windmill your arms. Flop your hands vigorously at the wrists. (5) When you arrive home, take a brisk walk or other exercise, get into a cold shower or hot bath and retire early. (6) The antidote for excess adrenalin in the blood is exercise. (7) Change your routine. Go to a different restaurant for lunch; walk instead of ride to the station. (8) Organize the pressure. Take time to analyze your job properly. (9) Learn to equalize your stresses. The human body, like the tire on a car, wears longest when it wears evenly. Get variety in your everyday life. (10) If you can't cope with your tensions, see your physician. Mottram Torre, "Emotional Tensions: Destructive or Constructive?" in Irwin, *ibid.*, p. 140.

3. The response of the electronics supervisors to stress is reported in F. L. W. Richardson, *Talk, Work, and Action* (Monograph no. 3, Society for Applied Anthropology, Ithaca, N.Y. 1961). The automobile assembly plant is described by Robert Guest, *Organizational Change—The Effect of Successful Leadership* (Irwin-Dorsey, Homewood, 1962).

4. On inadequate attention to time, one writer concluded that the most important discrepancy between our theoretical models of decision-making and the experience of some 35 American companies is the failure of the models to recognize that executive time is the scarcest resource. (Yair Aharoni, *The Foreign Investment Decision Process* (Harvard Business School, Boston, 1966). Also, in discussing the success and failure of corporate mergers, John Kitchey notes that lack of appreciation of the new demands on the time of executives of the acquiring company is one of the primary causes of failure. ["Why Do Mergers Miscarry?" *Harvard Business Review,* Vol. 45, No. 6 (November-December) 1967, p. 84]

5. Harry Levinson, a counseling psychologist to many executives, comments on this tendency to overwork in the ill-defined situation. (*Executive Stress,* Harper & Row, New York, 1970, p. 53). In their survey of the first years of the careers of graduates from Carnegie-Mellon Graduate School of Industrial Administration, Dill *et al.* point out that not knowing precisely what is expected and what are the criteria for performance are the most common difficulties of young managers. (William R. Dill, Thomas L. Hilton, Walter R. Reitman, *The New Managers—Patterns of Behavior and Development,* Prentice-Hall, Englewood Cliffs, 1962, p. 152)

6. These conclusions are from Carlson, *Executive Behavior.*

7. For a popular discussion of the "stimulus struggle" and man's need for a proper level of stimulation, see Desmond Morris, *The Human Zoo* (McGraw-Hill Book Co., 1969). Morris' description of how people and animals respond to inadequate stimulation comes very close to what seems to go on in organizations. He writes: "When a man is reaching retiring age he often dreams of sitting quietly in the sun. By relaxing and 'taking it easy' he hopes to stretch out an enjoyable old age. If he manages to fulfill his sun-sit dream, one thing is certain: he will not lengthen his life, he will shorten it. The reason is simple—he will have given up the Stimulus Struggle . . ." (p. 182).

8. See C. Northcoate Parkinson, *Parkinson's Law* (John Murray, London, 1958). An extreme example of excess time on the job is contained in a recent article: "Ever since man was sentenced to live by his sweat (Genesis, 3:19), he has been fascinated by the notion of an ever shorter work week. Six days, five days, four days—and say, how about no actual work at all? A return to Eden maybe? Well, no. Not according to 40 clerks of the Burlington Northern railroad, who are being paid around $30 a day for doing exactly nothing. In fact, a letter from one of them to The Chicago Sun-Times last week began: 'From Cell Block 101. Help!!'

"Their workless work week grew out of last year's merger of the Great Northern, Northern Pacific, Chicago, Burlington & Quincy, and Spokane, Portland & Seattle railroads into the 25,000-mile Burlington Northern system. As a result of a job-protection agreement with the Brotherhood of Railway & Airline Clerks, the new company found that it had 60 clerks in its Chicago office with nothing to do. At first the railroad tried sending them home early, but this stirred up resentment among the 700 other workers in the office. So it was decided a month ago that the redundant clerks should put in a full day's loafing for a full day's pay. They were ordered to report at 8 a.m. daily to a sparsely furnished room on the second floor, where a timekeeper dutifully checked them in, allowed them their ten-minute coffee breaks and a half hour for lunch and then checked them out at 4:40 in the afternoon. The 60, most of them women (about twenty have since been transferred to vacancies at the railroad's Cicero yards), spent their time reading newspapers, magazines and paperbacks, playing cards and checkers, knitting or just chatting. And they didn't like it.

"The first day," said one of the women clerks, "I thought, 'Hey, this is going to be all right.' But now I get out at night feeling more tired than I ever did in my life. I get home and drop exhausted." (*Newsweek*, May 10, 1971; copyright Newsweek, Inc.)

9. The comments by the commanding officer of the Maddox are reported in Frankel, *High on Foggy Bottom*, p. 93.

10. Evidence for higher satisfaction at higher levels is offered by Lyman Porter, "Where Is the Organization Man?"

11. Jay W. Forrester of M.I.T. writes: "Competitive position is often believed to rest on secrecy to a far greater extent than is the fact.

Information is withheld from individuals inside the organization on the excuse that this keeps information from outsiders. Secrecy is a poor foundation for success compared with competence, and to maintain secrecy reduces competence." ["A New Corporate Design," *Industrial Management Review,* Vol. 7, No. 1 (1965), p. 11.] In this article, Forrester discusses how emerging information technology can be utilized to expand informational access and organizational openness.

12. For a clear description of the use of fear for motivating managers in a famous corporation, see W. H. Rodgers, *Think: A History of the Watsons and I.B.M.* (Stein & Day, New York, 1969).

13. Estimate on windmilling advanced by Chris Argyris, *Interpersonal Competence and Organizational Effectiveness* (Irwin-Dorsey, Homewood, 1962), p. 274.

14. The T-group is from Argyris, *ibid.,* p. 186.

15. A classic study of prejudice is Theodore W. Adorno, Else Frenkel-Brunswik, Daniel J. Levinson, and R. Nevitt Sanford, *The Authoritarian Personality: Studies in Prejudice* (Harper and Row, New York, 1950).

16. The love of busyness is discussed by Arthur N. Turner and George F. F. Lombard, *Interpersonal Behavior and Administration* (The Free Press, New York, 1969), p. 585. A former chairman of Inland Steel writes about the self-appointed martyr that is the poor, harried, overworked manager: "Pity the overworked executive! Behind his paperwork ramparts, he struggles bravely with a seemingly superhuman load of responsibilities. Burdened with impossible assignments, beset by constant emergencies, he never has a chance to get organized. Pity him but recognize him for the dangerous liability that he is." (Clarence B. Randall, *The Folklore of Management,* Little, Brown and Co., Boston, 1961).

17. L. F. McCollum of Continental Oil Company, quoted in editors of *Nation's Business, Lessons of Leadership,* p. 128.

18. de Grazia, *Of Time, Work and Leisure,* p. 35. (© 1962, by The Twentieth Century Fund.)

6 Managing Time—Segmentation

The central message of most advice on time management is twofold: segmentation and concentration. Separate controllable from uncontrollable time (and convert some of the latter to the former if possible); consolidate and focus activities in controllable time. Simple as this message appears, it is difficult to implement because each tactic incurs costs as well as benefits.[1]

Long ago François Rabelais wrote: "How shall I be able to rule over others if I have not full power and command of myself?" This view underlies the contention that managers must actively control as much of their time as possible.[2] Implied is that one is not really a manager if he lets the job rule him instead of the reverse. If a manager lets events fully determine what he does, his ability, knowledge, and position will slip away because he is too busy "operating." Accordingly, he should distinguish between two kinds of time. On the one hand there is time that is not under his control—variously termed uncontrollable time, job-imposed time, required time, fixed time, and response time. The last term is most appropriate because this is the time consumed responding to events and people, with their requests, demands, and problems. On the other hand there is the time a manager can control—called controllable time, self-imposed time, disposable time, and discretionary time. Again, the last term is most appropriate.[3]

Peter Drucker suggests that no manager has more than 25 percent of his time under his control. Others indicate that the controllability factor may run from 20 to 50 percent. Regardless of what it is precisely, most advisers contend that it should be larger. A well-known American Management Association film maintains that the difference between effective and ineffective executives is the number, significance, and rate of completion of their discretionary tasks, not how well they perform during response time.[4] As will be discussed later, we take exception with this downgrading of response time; it is important. Nonetheless, most managers would like to expand discretionary time. What are the techniques available?[5] Insulation, isolation, delegation, and simplification stand out.

INSULATION

Insulation consists of buffering the manager from organizational demands. The most common device is by scanning and sorting incoming communications, of which a classic illustration is the United States Presidency. By executive order in 1939, Franklin D. Roosevelt created the Executive Office of the President "to protect the President's time" by "excluding any matter that can be settled elsewhere." This mechanism evolved into a group of assistants and counselors who stand between the President and almost all incoming communications. They decide: (1) what; if a matter should be brought to the President's attention, (2) who; by the originator or by a White House aide, (3) how; by letter or verbal communication, and (4) when.[6]

In addition to controlling incoming communications, these assistants can also organize them by grouping conferences to specific periods of the day which the Chief Executive prefers. Such grouping also aids the President to prime for meetings by concentrating his preparation on overlapping areas.

Of course, most managers do not have such elaborate staffs. To gain similar benefits from an insulating buffer, they must depend upon an assistant or secretary—a secretary that does much more than type and answer the phone. A competent buffer should have enough discretion to redirect about half of a manager's incoming messages to more appropriate people. On matters that should reach the manager, an astute subordinate can anticipate his needs and save time by gathering relevant data and preparing preliminary suggestions. The intent of such activity is to ensure that the manager reads each incoming communication only once before a reply is drafted.

Using a buffer to limit access and interruptions is widespread. It defends the executive against those who would manipulate his time to their own interests. Yet, for all its substantial benefits, the buffer mechanism has unfortunate defects. With the United States Presidency, buffering has been criticized as dangerous. The assistants may insulate the President from information they consider trivial, but which may really be important. The information reaching the President may be so filtered and biased that he is out of touch with reality. Former Vice-President Humphrey has commented that he never perceivd the magnitude of anti-Vietnam War feeling until he left office. And former White House Press Secretary George Reedy argues in *The Twilight of the Presidency* that the problem of keeping a Chief Executive in touch with reality is so severe that it poses a danger to the future of our country. Reedy maintains that the White House is enveloped in at "velvet cocoon."[7] Clark Clifford, former Secretary of Defense and a perennial adviser to postwar Democratic administrations, remembers an Eisenhower aide telling him that Ike was spared night work because his staff boiled 150-page memoranda down to two pages. Clifford replied, "There is only one trouble. If I could be the fellow who prepares the two page memo, I'd be President instead of Ike."[8]

Similarly, if less dramatically, every executive runs a risk of being isolated from his organization. Clarence Randall, a former chairman of Inland Steel, wrote about the loneliness and isolation of his post in his *The Folklore of Management,* where he disparaged the stereotype of the president being on top of every issue in his corporation. Many managers fear compounding this problem by overly rigorous buffering by a zealous subordinate.[9]

The maverick President Robert Townsend describes how he dealt with this decisively: he fired his secretary.[10] To control incoming calls, he simply informed the switchboard to connect calls only between 11–12 and 4–5. At other times, they were to take names and indicate that the president would call back. Of course this practice introduces a time delay, but Townsend maintains that few matters can be urgent within four hours. The practice also groups all telephone communications into small portions of the day. In addition, without a secretary, no one sits outside the president's office blocking entry. Without this buffer, perhaps this chief executive can promote closer relations with his subordinates by ensuring that everyone gains access, not just the most energetic and aggressive subordinates. Still, operating without any buffer seems extreme. More modestly, another company's entire management group agreed to shut down the telephones for one hour in the morning, allowing the switchboard operator to take messages during that period rather than putting through any incoming calls.

ISOLATION

The most definite form of separation is physical isolation from the normal workplace. Across the street from the White House in the Executive Office Building, President Nixon has a small private office where he can work alone. All Presidents have found it essential to have secluded retreats as in Hyde Park, Key West, and Camp David. Business executives may also maintain two offices (an entrepreneur in Fairfield County, Connecticut, recently erected an office building specifically intended to provide second offices for Manhattan executives residing in Connecticut), but the more common practices are to work at home one day a week, close the office door and refuse all calls, or merely to utilize travel time. "You don't have to talk to anybody on a plane," comments a company president. "A couple of hours on a plane and you can do a tremendous amount of concentration, and that's impossible in the office. You have to do it at home or under other conditions."

What is involved here is an effort to consolidate chunks of uninterrupted time in order to think more deeply about organizational problems. Drucker tells of a bank executive who maintains that, for ninety minutes, "My secretary has strict instructions not to put anyone through except the President of the United States and my wife. The President very rarely calls and my wife knows better."[11] Such chunks of thinking time are a great advantage over the short ten-minute intervals mentioned in Chapter 5. But are they long enough?

At the most modest level, a manager may gain a little isolation by refusing to participate in one of the most insistent demands, the working luncheon meeting. There he eats into his free time on an expense account at a lower rate of remuneration than he receives for his own time. Instead of thus working for their lunch money, some executives eat alone in the office, sit in the park, or go to the zoo (medical research supports skipping the working lunch in favor of eating alone because it is one of the most stressful executive activities).[12]

Note that partial isolation is not postponing routine or boring tasks to the evening. Sometimes managers must take work home of course, but putting things off until evening is often more procrastination than time management; it enables them to avoid organizing their time during the day.

Periodic isolation seems so simple, but risks are involved. Can the manager afford to be unreachable? The President of the United States clearly cannot, so he has extensive communications equipment travel with him. Many managers take much the same view of themselves. If they are accessible, however, how can they confine interruptions only to true emergencies? Who makes this determination?

Decision about withdrawal frequency and length will depend upon position demands, the magnitude of potential emergencies, and faith in subordinates. An operating supervisor can be out of touch for only short periods; the demands of his technology and subordinates are too insistent, too quick-paced. A high-level general executive will have much more leeway.[13]

Any withdrawal may create personal animosity. The associate or subordinate who calls with a matter that is important to him may be offended when he is denied entry. Even tougher, courage is required to withdraw from a superior's initiations for *any* length of time. He may expect his subordinate to be always on tap. In addition, the culture of the institution may discourage periodic withdrawal. At a meeting of church officials, one bishop maintained that withdrawal was simply impossible. The human values implicit in his religion's mission so permeate church administration that the administrator is expected to be a shepherd and servant to his associates and subordinates. The value system requires that he be available, understanding, and responsive. This may well hinder the performance of the institution. Even in business some managers are afraid to close their doors—afraid that a closed door may suggest that some secret changes are being plotted, or because of a misplaced pride that "my door is always open," or maybe even that his boss will think that he is asleep.

Superiors should recognize their subordinates' problems in this regard. Managers rarely are sensitive about their responsibility to protect their subordinates' time for thinking, thereby contributing to the impulse to act at the expense of thought. Young managers particularly are caught in a tension between reflection and production.

A superior should protect his subordinates by requiring thinking time and refusing to let them equate busywork with deliberation.[14]

Some managers fear isolation and withdrawal because of their own personalities. Perhaps they fear the ambiguity of such periods; perhaps to maintain their drive, they must be on their feet talking with others; perhaps they fear time for thought because they have no idea of what to think about.[15]

DELEGATION

Delegation often begins with pruning; a manager examines his job and asks himself a negative question:[16] *"What am I doing that does not need to be done by anyone?"* The wisdom of hindsight may pinpoint recurring behavior that only makes minimal contribution to the organization. Would elimination adversely affect performance? Some kind of historical record such as a central inventory of required reports is valuable for this examination. By periodically considering the record, a manager can determine what reports are obsolete because conditions have changed. For example, many years ago a photographic film company initiated a safety report on motion picture film then manufactured of highly flammable nitrate base. The purpose of the report was to keep track of the film because it became more dangerous with age. Yet the film was being traced and the report compiled on new film thirty years after the industry had converted to nonflammable acetate material. As shall be discussed further in Chapter 11, no organization is free from its history. The decisions and actions of yesterday limit and influence present behavior, but a manager can attempt to curtail his servitude to the past by eliminating inherited tasks that no longer contribute.[17]

In pruning, a second question is *"What could be handled as well by others?"* Most managers tend to exaggerate their own importance, but one's importance may be even greater if personal time is focused and distractions are reduced by delegating to others.

Delegation has its origins in the overworked owner-manager who just couldn't keep up with everything. Therefore he hired someone. Although he didn't give his new employee much authority or autonomy, he had the subordinate assume some of his activities. Much delegation in large organizations shares this general theme: begrudging assignment of tasks that the manager would prefer to do himself if only he had the time. And all is not saved when a manager delegates. For now he must supervise, monitor, and correct the subordinate, especially because the new man initially is not as competent. Many new supervisors feel that they can do the job faster without a subordinate; in the short run, they are probably right.

Screening incoming communications, bunching similar activities, simplifying, and so on are all ways of doing *more* in a fixed time. Delegation is different; it is a way of *doing* less. It saves time for the delegating manager, but it does not save time for the organization. *Delegation always uses more total time.* The delegator's intention is to free himself for exploration of different and new ideas so that the organization accomplishes *more work in more time.*

Here are some guidelines that may provide help in thinking about delegation. In general, a manager should:

(1) Retain activities in which he enjoys the greatest comparative advantage over subordinates—even if these are not the tasks at which the manager is most absolutely effective.

(2) Retain activities concerned with change; delegate those primarily intended to maintain stability. The latter are more programmable so that other people may be able to handle them. Change is more difficult.

(3) Retain discretionary activities, delegate uncontrollable response activities. Responding to periodic problems, answering most mail, conducting routine meetings are more feasible to delegate than the thinking, planning, and innovating of discretionary time.

In order to avoid needless expenditure of time on reports and checks, it is important to realize that delegation is not a "go, no-go" process. There are degrees of delegation determined not so much by what is delegated, but what the subordinate is to do *after* he performs the work. For example, typical delegation patterns are:[18]

(1) Take action—no further contact with superior necessary.

(2) Take action—let superior know what you did.

(3) Look into this problem—let your superior know what you intend to do; do it unless he says not to.

(4) Look into this problem—let superior know what you intend to do; delay action until superior approves.

(5) Look into this problem—let superior know alternative actions available with pros and cons; recommend one for superior's approval.

(6) Look into this problem—give superior all the facts; superior will decide what to do.

When delegating or accepting a delegated task, a manager should ensure that he knows what pattern he is working under. It is seldom clear from the initial instructions. Clarity will save much time and embarrassment. After the pattern is defined, the superior should be consistent in sticking to his intentions through to completion of the delegated task. Otherwise, subordinates will tend toward patterns 4–6 in fear of taking action for which they will be criticized.

Delegation is obviously a logical step when the demands on a manager are great (especially if a subordinate has too little to do). Yet, because of fear of losing status and control, its time saving possibilities are too infrequently realized.

SIMPLIFICATION

Managing is not a simple task; no tricks exist for uncovering a rich lode of untapped time. Yet simplification of personal communications may speed up the managerial process, saving some time. In total, these communication techniques are probably less important than insulation, isolation, and delegation, but they are helpful.

Grouping outgoing communications is a basic tactic that many managers arrive at through experience. Common sense suggests grouping the dictation and drafting of outgoing correspondence and telephone calls into a convenient part of the day. One of the reported advantages of not accepting incoming calls immediately is that a manager can return them to suit his own schedule. Otherwise inviting others to call invites interruption. In addition, the initiator of a call may be able to maintain control of the conversation more effectively.

Handling correspondence only once also makes sense. All too frequently incoming communications are read by superior, subordinates, and superior again before a reply goes out. Immediate response to incoming communications is another common managerial practice for saving time. Responding immediately is more self-discipline than a technique. A manager should approach his incoming communications expecting to respond immediately rather than automatically dropping everything into the pending file where it must be reread later before replying.

When initiating communications, sometimes a manager has a choice between writing or talking. Research suggests that the typical manager prefers verbal communication.[19] It is difficult, however, to judge which is more efficient. President Nixon is said to dislike the "laying on of tongues," as White House locution has it, and much prefers spending three minutes reading a memo to hearing someone out in person for a quarter of an hour. Conversations are difficult to keep under control, of course, but they do allow more flexibility and promote two-way communications which hopefully lead to better understanding. Here are some guidelines for choice between writing and talking:

(1) Does the manager want any questions about his communication? If not, a one-way announcement by letter or memo is best. Verbal announcement at a large meeting will also avoid questioning if he indicates no desire for the audience to speak up. Small groups and individuals are frequently more difficult to control in this respect.

(2) Does the manager want the socializing that tends to occur with verbal, face-to-face communication? He should consider this. One adviser on time management suggests reducing social content from conversation and sticking to the topic. Of course some discipline is necessary, but to be exclusively business in all discussions is offensive and a hindrance to group effectiveness.

(3) Does the manager want to make himself available so that it is easier for the other party to raise different matters? If yes, go talk. This is an important decision. It runs counter to close management of time, but the manager may gain advanced information that he might never obtain except by just being there.

Once again, behavior choice depends on the job. High-pressure, low-autonomy, short-time-span positions force a manager to utilize one-way verbal or written messages. Research with higher executives, however, indicates that they tend to utilize two-person talks more than group meetings.[20] The face-to-face situation facilitates two-way communication and the introduction of unscheduled topics. In the short run, this two-person style may be more time-consuming because meetings and printed notices offer greater efficiency in conveying information to many people. Nonetheless, the more intimate style may save time eventually through increased motivation and understanding which will allow greater delegation.

Some people, like President Nixon apparently, prefer reading to listening or discussing. To choose an inappropriate method for a particular person may necessitate repeating the message in a more desired mode. President Kennedy reportedly found the Department of State unresponsive to his telephone calls even though he was the Chief Executive. Their system was geared to written communications, not verbal. J. F. K. learned to send memos whenever he wanted a response.[21]

A manager may save some time even when people come to speak to him. The crude way is to set a time limit upon the visitor, or to act bored, look at the clock, and shuffle through papers. Such mechanisms will probably be effective with subordinates and associates; they will leave quickly, but they may also be discouraged about coming in at all. Discussions may be more efficient, however, if the manager requests that his subordinates complete certain steps before coming—such as checking records, communicating with concerned others, and estimating how the matter under consideration will affect management time. Gathering such information before the meeting may allow the issue to be settled quickly without later repetitive discussion.

On written communications from subordinates, a manager might announce that since he intends to read everything, only important material should be submitted. Similarly, some executives prefer that staff specialists announce a recommendation rather than just outlining alternatives. Responding to one recommendation will utilize less time, but a manager runs a risk of losing control to his specialists. President Eisenhower was said to insist upon "completed staff work" by which he meant the submission of a plan that only required his approval for activation.[22] In contrast, President Kennedy (especially after the Bay of Pigs debacle) wanted to be involved with the staff's discussion of alternatives; they were not to make a recommendation unless asked to.

The intent of these techniques is to reduce and confine time devoted to activities over which managers exercise little control. In the next chapter we shall consider utilization of controllable time.

REFERENCES

1. Originally, our intention was to discuss "how not to save time" followed by "correct ways to save time." However, it became evident that this was a false dichotomy. Some ways of managing time are appropriate for certain positions, wrong for others. Each technique has its costs and benefits. Therefore, one must consider the various possibilities, including advantages and disadvantages.
2. Long-time Harvard Business School professor Robert Merry states it succinctly (and idealistically): "You must decide for yourself how much time you will plan to spend on what." ("Pressures on the Presidency," mimeo, Institute for College and University Administrators, Boston, June 30, 1965)
3. Auren Uris distinguishes between three kinds of time: fixed, semi-fixed, and variable. His suggestion is that these be balanced. The distinction seems unrealistic to me. Uris' article is short, modest, and helpful, but the title is misleading and typical of the inflated promises and oversell in the time management area. The article: "How to Have Time for Everything," *Duns Review,* August 1957, p. 53. Some books on managing time are: Joseph D. Cooper, *How to Get More Done in Less Time* (Doubleday & Co., New York, 1971); Drucker, *The Effective Executive* [1]; Engstrom and Mackenzie, *Managing Your Time;* Eldwin Feldman, *How to Use Your Time to Get Things Done,* (Frederick Fell, New York, 1968); Carl Heyel, *Organizing Your Job in Management* (American Management Association, New York, 1960); Godfrey M. Lebhar, *The Use of Time* (Chain Store Publishing Corp., New York, 1968); Raymond O. Loen, *Manage More by Doing Less* (McGraw-Hill Book Co., New York, 1971); Philip Marvin, *Multiplying Managerial Effectiveness* (American Management Association, New York, 1971); Auren Uris, *The Executive Deskbook* (Van Nostrand Reinhold, New York, 1970).
4. In his film, "Managing a Manager's Time," William Oncken divides executive duties into three categories: job-imposed; system (organization)- imposed; and self-imposed. He suggests that what sets apart successful executives is the number, significance, and rate of completion of their self-imposed tasks. These are tasks not called for in the job description and not required by the organization, but voluntarily assumed by the executive who sees a need and a means of fulfilling it on his own. (American Management Association, 1961)
5. One survey of executives reports their "greatest time-savers": (1) the committee of two—avoid involving unnecessary people in the decision-making; (2) correspondence-fast answers; (3) correspondence—handle only once—don't put it back in the pile; (4) correspondence follow-up file—if you must wait for information before answering, mark for FU file. Secretary pulls out in one week and returns to you with infor-

mation necessary for answers; (5) insure understanding when delegating; (6) appointments by secretary; (7) delegate reading; (8) conference phone call with prearranged agenda; (9) have secretary answer correspondence; (10) shorter memos and letters; (11) wastebasketry; (12) form letters; (13) planning and organizing time; (14) a good secretary; (15) management training. (Engstrom and Mackenzie, *op.cit.,* p. 224).

6. For an interesting discussion of various Presidents and their styles, see Clinton Rossiter, *The American Presidency* (rev. ed, Harcourt, Brace and World, New York, 1960).

7. On the dangers of overinsulation, Russell Baker, the political satirist of the *New York Times,* writes of the tough and brilliant but mythical P. C. Biederbaum whose central task is "to husband the President's precious time, and he does it with a zeal that often infuriates persons who yearn to consume that most precious time. . . . Ambassadors, Senators, and other staff aides, and Mrs. Nixon complain bitterly about Biederbaum's refusal to grant them time with the President. The President himself has been heard to mutter wistfully that he wished Biederbaum would 'let me in on what's happening in Vietnam.' Biederbaum, young, brilliant, hard-nosed, ruthlessly efficient, creaking with clichés, is unmoved. He believes that his first task is to protect the President. For the past two weeks he has even refused to grant a moment of the President's precious time to Biederbaum. Mr. Nixon reportedly has tried to complain, but Biederbaum refuses to grant an audience to anybody whom the President can complain to." (*New York Times,* August 12, 1969; © 1969 by The New York Times Company. Reprinted by permission.

8. Clark Clifford is quoted in *Time,* June 8, 1970.

9. Engstrom and Mackenzie observe that the isolation of the top executive can be helpful: "If the chief executive should be spending from ⅓ to ½ of his time planning ahead, how better can this be done than by taking advantage of the relative isolation at the top? The most successful executives go to great lengths to set aside and preserve such time, safeguarded from interruptions and the demands of the routine." (*Op. cit.,* p. 236)

10. Robert Townsend reports on how he fired his secretary in *Up the Organization,* p. 168.

11. The executive who allows only the President or his wife to call is quoted by Drucker, *op. cit* [1].

12. See Irwin, *What the Executive Should Know about Tensions.*

13. On chief executive absence, Sune Carlson writes: "There were chief executives in our study who had organized the management of their firms in such a manner that they could be away a large part of their working time without any major difficulties arising. The management of these firms was characterized by a far-reaching decentralization of authority and responsibility, and by the fact that a definite person or committee of persons was appointed to deputize permanently for the managing director when he was away. In other firms, the chief executive might be absent from the office much less and nevertheless there were definite signs of a "vacuum" in the top-management. There was irritation among subordinates because of the lack of contact with the chief executive and because of the delays in decision making." (*Executive Behavior,* Strombergs, Stockholm, 1951, p. 66)

14. The responsibility for senior managers to be concerned about the use of time by their younger subordinates, and their protective role, is discussed very well by Harry Levinson in *The Exceptional Executive* (Harvard University Press, Cambridge, 1968), pp. 160 ff.

15. On different patterns of withdrawal and talking, a manager critically compares two superiors: "Whereas Arnold has been able to sit down and lock the door and work at his desk for hours at a time on a realistic and detailed sales plan, Bart would get upon his feet and talk and talk and contradict himself. The office joke was a definition of his management policy, namely, the idea he just got until he has lunch with someone else tomorrow." (Quoted by management consultant Chester Burger, *Executives Under Fire,* The Macmillan Co., New York, 1966), p. 133. (Copyright © 1966 by Chester Burger.)

16. The negative questions are from Peter Drucker, "How to Manage Your Time," [2] *Harper's Magazine,* December, 1966, pp. 56 ff.

17. Drucker calls this "sloughing off yesterday." He writes: "This means that every executive forever has to spend time, energy, and ingenuity on patching up or bailing out the actions and decisions of yesterday, whether his own or those of his predecessors. In fact, this always takes up more hours of his day than any other task . . . But one can at least try to limit one's servitude to the past by cutting out those inherited tasks that have ceased to promise results." (*Op. cit.* [1], p. 104)

18. These patterns of delegation are from Harvey Sherman, *It All Depends: A Pragmatic Approach to Organization* (University of Alabama Press, Tuscaloosa, 1966), p. 83. © 1966 by The University of Alabama Press. Used by permission.

19. For fifty-four managers, the mean figures are:
7.9 hr/wk on written comunications received.
4.2 hr/wk initiating written comunications.
10.1 hr/wk on oral comunications started by others.
15.8 hr/wk on oral comunications self-initiated.

20. For eleven high executives, the mean figures are:
15.0 hr/wk in two-person talks.
10.4 hr/wk in small group meetings (5 or less people).
3.9 hr/wk in large meetings (more than 5 people).

21. Arthur M. Schlesinger, Jr., discussed President Kennedy's problems with the State Department in *A Thousand Days* (Houghton Mifflin Co., Boston, 1965).

22. In their study of young managers, Dill and his associates report that older managers rated highly those subordinates who submitted work that the superior could simply pass on without taking time to rework. Thus, managers highly value those who make the least demands on their time. This is understandable, but it does leave the superior vulnerable. Dill, Hilton, and Reitman, *The New Managers,* p. 152.

7 Managing Time—Concentration

Concentration pertains to discretionary time. Decisions on concentration affect response time, but only indirectly by narrowing matters brought to the manager. Concentrating thought and action is more difficult than separating discretionary time from response time or even accumulating substantial chunks of uninterrupted time. There are just no relatively clear-cut techniques such as insulation, isolation, delegation, and communication simplification.

The advantages of focusing, however, are great. By clearly defining and limiting the business, its objectives, and strategy, a manager reduces the enormous time wastage of needless forays into irrelevant areas. He can save much personal time by imposing a framework on information collection, for he will realize that attention need not be devoted to everything, merely those issues pertinent to his concentration. But, how difficult it is! For concentrating requires self-discipline and the courage to ask fundamental and critical questions about values, mission, and strategy. Accumulating chunks of time by working at home or isolating himself in his office is helpful for gaining time to think about concentration, but like the proverbial horse, a manager will not necessarily do it when brought to the water.

SETTING PRIORITIES

"I don't have enough time," managers protest. What they really mean is that the problem in question is not sufficiently important to warrant *taking* time for it. When there are more problems than time, decisions are necessary. These priority decisions will be made as a result of situational pressures or freely by the manager. If situational demands are allowed to predominate, important tasks may well be ignored. Therefore, a primary means of concentrating discretionary time involves setting priorities on the work to be accomplished.[1]

Setting priorities allows management to consider problems sequentially rather than simultaneously.[2] President Nixon is said to prefer a system that frees him to concentrate, mostly alone, on one big question at a time.[3] Consistent with Mr. Nixon's style, Peter Drucker also supports this concentration of "first things first, and only one at a time. This is the 'secret' of those people," he writes, "who do so many things and apparently do many difficult things. They do only one at a time. As a result, they need much less time in the end than the rest of us." Drucker goes on to suggest that effective executives concentrate in three ways:[4]

(1) They build on their strengths and those of their colleagues. This means concentrating on those activities that they do best.

(2) They concentrate on outward contribution. Results guide their efforts rather than just work. They begin by asking themselves what results are expected of them, rather than with the tools or techniques of the task.

(3) They concentrate on the few major areas where superior performance will produce outstanding results.

Investigation of technical research personnel tends to support these observations. Thus, the following factors seem to enhance development of new ideas: concentrating over 50 percent of one's time on one project, some isolation from colleagues in the same institution, but more communication with one's administrative superior and people outside the laboratory.[5] But most managers do not operate under conditions similar to research, where time is more controllable. Therefore, they find it difficult to be so rational or methodical. While contrasting his former work as a Harvard professor with his job as Presidential adviser, Henry Kissinger stated the essential problem: "Before, I could always deal with a problem in what I thought was a reasonable way. I could work at something until I thought I understood it. Here, I've had to put problems in order of priority and hope that the low priority ones won't hit us before we get to them."[6]

A former Chairman of New Jersey Standard Oil defines the problem similarly:

> *Question:* "Have you found in your years in business that there is a best time to think and plan your day?"
>
> *Answer:* "No, I don't think so. As a matter of fact, in our business, we might plan this way, but the plan won't last very long. You can walk in the office at 9:00 in the morning and at 9:15 something you didn't have any idea about will rise in the way of a pretty important problem or crisis. So what you do is you give priority to the things on the basis of your judgement of what has the highest priority."[7]

But what are the specific priority criteria that determine what matters are dealt with? Three are apparent: order, urgency, and contribution.

ORDER OF ARRIVAL

Perhaps more than in any other culture, Americans do first things first. Arrival time is a criterion for obtaining a seat at the theater, restaurant, or on a bus while in many countries age, sex, or status are more relevant criteria than first come, first served. Priority due to order of arrival is very democratic, of course, and it is a primary criterion during a manager's response time, but it proves poor as a method for allocating discretionary time. Lesser matters will outnumber and submerge more important ones if handled in order of occurrence.

URGENCY

A sense of urgency stems from promixity of a task deadline and fear of not meeting it. Managers tend to give great weight to urgency. For example, virtually every commercial desk book or pocket planner has a column entitled "to be done today" with appropriate spaces for marking completion. Such planners are invaluable for keeping track of where a manager is supposed to be when. As logical as such a system seems, however, it is really not planning but rather response to urgent problems. Presumed is that first thing in the morning a manager will preview his day's activities and list time for the most important ones. The weakness of this approach is that urgent "quick-time" problems tend to be listed and less urgent "slow-time" tasks left out.

This tendency to work on quick jobs is seducing. A manager sits down at his desk; small tasks and big problems are queued up. Many men start on a few quick jobs (regardless of importance) in order to ease into the day's work. This is not bad in itself, but it can be extended too long, so that a manager deludes himself. For a manager with high achievement need who seeks and enjoys concrete feedback on completed tasks, to remain on quick jobs is attractive. They roll by giving an illusion of progress. Like the proverbial poor, however, the small, urgent-appearing tasks are always with him. They seem to flow in so that he never gets to the slower and often more important jobs. As President Eisenhower arranged his affairs so that only the truly important and urgent matters came across his desk, he reportedly discovered that the two seldom went together. He found that the really important matters were seldom urgent, and that the most urgent matters were seldom important.[8]

Fear and a sense of urgency may extend beyond oneself to others. Yet behavior may still be misdirected. In the electrical equipment price conspiracy trial of the early 1960s, some managers indicated that they balanced three fears: first, fear of not reaching a sales quota, not qualifying for a bonus, etc.; second, fear of being caught violating the antitrust laws; and third, fear for their workers if pure competition existed in an overcapacity and cyclical industry. The third fear tipped them toward collaboration with competitors.[9]

Regardless of whether fear is self-centered or other-concerned, it tends to be a short-run criterion. However understandable, price conspiracy and market sharing was only a temporary expendiency. Time was diverted from efforts to find a better long-lasting solution to the industry's problems. At times fear will and should dominate—as when a military unit is under attack, a student riot is imminent, or bankruptcy looms. In such situations, the present must take precedence over the future. Yet, if urgency predominates for very long as the criterion for assigning managerial time, failure can be expected.

Segmentation is a way of handling this problem. So much time for quick jobs (perhaps sandwiched into gaps during response time); so much time for slow jobs. This means reserving a block of time in that pocket planner for a long-range problem just as if it had to be solved today, even though the manager knows it does not and it will not. *No other point on managing time is more important than this.* Segmenting quick and slow jobs is probably the most common device among effective time managers. They rigorously reserve a portion of time periodically (perhaps daily, more likely weekly or biweekly) for important, if less urgent, tasks. During this time, urgent matters are buffered and postponed (except, of course, items of overwhelming urgency). To facilitate this concentration, managers may isolate themselves. Obviously, substantial self-discipline is necessary for such a system.

CONTRIBUTION

This priority criterion is measured by contribution to the organization, whether profits, return on investment, sales, service, or whatever the relevant objective. The observation that effective managers concentrate their effort where their contribution will be greatest implies that they determine what activities have the greatest potential return. Thus time is treated as what it is: a resource to be invested.

Time as an Investment

Curtis Jones of Harvard Business School maintains that: "The minimization of demands on executive time deserves almost equal attention to that which businessmen now devote to profit maximization." He offers the following illustration.[10]

Rankings of Hypothetical Projects by Return on Executive Time (dollar figures in millions)

Project	*Return on investment*	*Capital req'd*	*Annual return*	*Net gain after cost of capital*	*Man years req'd*	*Return per man-yr*
A	50%	$10	$5.0	$3.0	3	$1.00
B	40	6	2.4	1.2	3	.40
C	35	15	4.5	1.5	2	.75
D	30	10	3.0	1.0	1	1.00
E	20	15	3.0	0	1	0

This is a hypothetical company that has 20 percent cost of capital and six "executive years" available to carry out a proposed list of five projects. Consider:

(1) If the company does not recognize the executive time limitation in advance, it may select projects A, B, C, and D and then run into severe trouble when it runs out of executive time to manage the projects.
(2) If the company is sufficiently well managed to plan the demands on executive time, it may select only projects A and B, for a total investment of $16 million and a net gain of $4.2 million above the $3.2 million cost of capital.
(3) If the company ranks the projects by their return on executive time, it may select projects A, C. and D for a total investment of $35 million and a net gain of $5.5 million above the $7 million cost of new capital.

Note that no novel accounting is required; it is more a style of thinking than a rational technique. In utilizing discretionary time, managers should train themselves to think in terms of investment and return, concentrating their time on the most promising. Many managers say that this is more subjective than rigorously rational—perhaps even requiring faith in possible returns that are difficult for others to see.

Time and Multiple Objectives

The example above implies that management has relatively few objectives by which contribution is measured: return on investment, capital required, gross time required, annual return, etc.—all financial or numerical measures. In reality, of course, organizational objectives are more complex, so management must allocate time according to multiple objectives and ambiguous priorities. Concentration of time is worse than useless if management has no sense of the multiplicity of its objectives and their relative importance.

Awareness that profits or even net return on investment are not the only objectives of a business is widespread through the persuasiveness of Peter Drucker, Rensis Likert, Douglas MacGregor, Harold Simon, and others less familiar to practicing managers. Too many variables and external events affect profits for these to be a sufficient guide to most managerial behavior and decision-making. Allocating time according to supposed contribution to profits tends to reflect mainly urgency and not the long-range health and success of the organization.

Following is a composite list of objectives areas that should be of concern to all businesses.[11] They are organized in random order in two columns: (1) emphasizing results objectives—ends, and (2) emphasizing process objectives—means. The distinction is a little arbitrary, of course, because one shades into the other, but time must be allocated to the second-column objectives or the first will eventually not be achieved.

(1) Results Objectives	(2) Process Objectives
Market Standing	State of physical and financial resources
Productivity	Identification (achieving clarity, consensus, and commitment to organizational goals)
Profitability	Integration (integrating individual needs and organizational goals)
Social Contribution/public responsibility	Collaboration (producing mechanisms for the control of conflict)
Innovation	Stabilization/adaptation (responding appropriately to changes induced by the environment)
Service	Development and revitalization (dealing with growth and decay)

The items are not in order of importance because it is doubtful that one order applies to all organizations and units. It depends upon the circumstances of the relevant unit. For example, here are three lists of objectives arranged in order of importance by managers/participants in each:

A. Objectives for a Bank (as determined by its managers)
(1) customer service, (2) organization—development of quality manpower, (3) employee relationships, (4) innovation, (5) recognition, (6) growth, (7) immediate earnings.

B. Objectives for a home products manufacturing company (as determined by its executives):

(1) Objectives for customers—highest quality products, decreasing prices, and accurate and rapid service.
(2) Objectives for employees—job security, fair wages, just management, opportunity for advancement, individual consideration.
(3) Objectives for managers—opportunity for education and growth.
(4) Objectives for communities—be good citizens, support charity, pay fair taxes, promote civic improvement.
(5) Objectives for stockholders—sound profit, research/development/introduction of new products, plant maintained and expanded, fair dividends.

Objectives for Engineering Department of Power Company

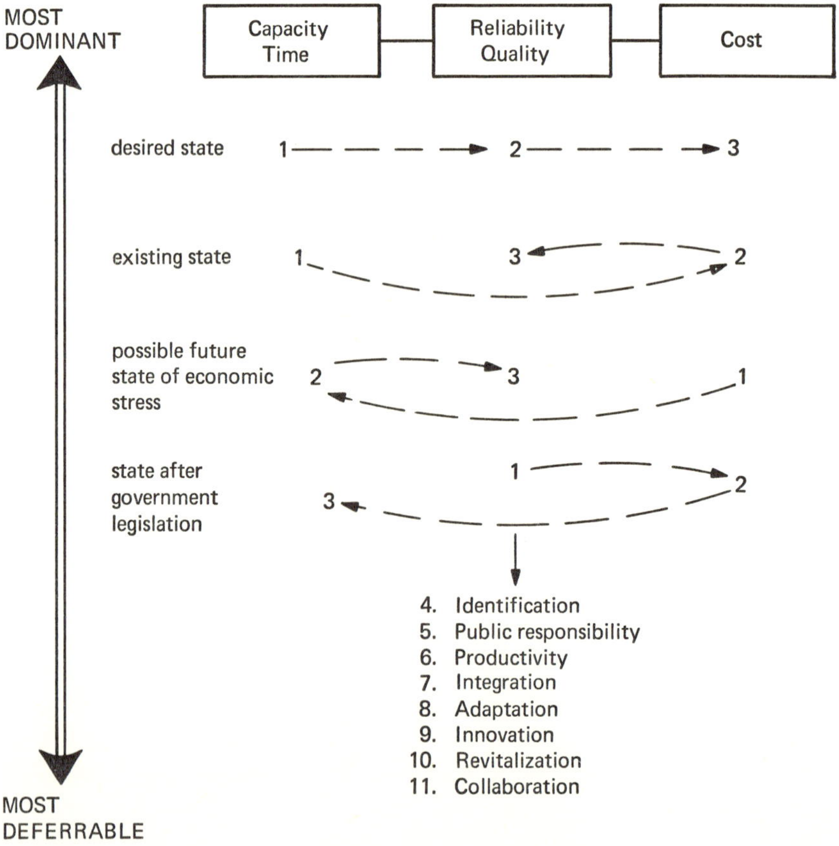

C. Objectives for a university (as determined by a group of faculty and administrators):
(1) service to constituencies, (2) development of staff, (3) control/minimize expenses, (4) innovation, development and dissemination of new knowledge, (5) employee relationships, (6) growth, and (7) recognition.

Now let us consider another institution in greater detail. The diagram on page 80 shows how the fifteen man executive team of the Engineering and Research Department of a large private power company organized the Department's objectives according to their priority system.

This list was developed only after several hours of free discussion among the management group. It is not the same as their initial list; that seemed to be more idealistic. Initially, the managers (all engineers) stressed integration and innovation. Later discussion, however, was centered on capacity/time/reliability/cost. Most meeting time was concerned with these matters and relatively little on-the-job time was devoted to integration or innovation. (One executive finally concluded that the Engineering and Research Division really did not want innovation after all! They needed adaptation and adoption of ideas developed elsewhere.) This is not to denigrate the importance of the more deferrable objectives, but to suggest that management perceived them as less urgent.

Allocation of Time to Multiple Objectives

Given a similar analysis of objectives, any management must ask itself two questions:

(1) Are we allocating time in accordance with our defined priorities?
(2) Are we allocating sufficient time to more deferrable, but still important objectives?

The second question is particularly important because setting priorities is actually more critical for deferrable objectives. Time probably will be allocated to the more dominant objectives because of urgency. Managers will be unable to avoid them. But, process-oriented, more deferrable objectives also need time or the dominant matters eventually will suffer. The primary benefit of the lengthy discussion in the power company was to pinpoint just what objectives were being slighted.

How much time should be devoted to each objective? A definitive answer is impossible because situations differ and so little research exists that details time allocation for managers in these different conditions. A recent survey suggests, however, that top-level executives at least are becoming increasingly engaged in innovation, integration, and development. This seems to be the main conclusion from extensive interviews in large corporations. Here are comments by four executives:[12]

> #1 "I spend a substantial portion of my time in selecting and moving people. We expect to expand our rotational system and this will take more of my time.
>
> #2 "Certainly fifteen per cent, I think, of the time of most of the higher executives is spent on development of people, planning for their future development."
>
> #3 "It's my number one job. I spend more time on selection and development of personnel than on any other activity."
>
> #4 "Top executives seem to have a consensus that more of their time will be devoted to coordinating research and development, developing and implementing new products."

As suggested in Chapter 3, we should be a little skeptical about what managers say they do; they have a tendency to overestimate time allocated to activities that they actually are neglecting. Nonetheless, as a general guide, it appears that effective managers devote more time to more deferrable objectives. For high-level executives, this might be up to one-half of their time, for middle managers, up to one-fourth, and for operating supervisors perhaps only one-tenth of their time. Many practicing managers will protest that this is too low, that they spend more than twenty-five percent of their time on deferrable matters. This is unlikely. If they would arrange their objectives in a fashion similar to the power company example, it is probable that they are not devoting anywhere near one-fourth of their time to the lower half of their objectives list. Although it is impossible to prove at present, they should be.

Regardless of the specific percentage of time, the central point is that time must be allocated to more deferrable objectives if the organization is to survive in the long run. The most popular technique for ensuring that time is utilized in this way is to separate functions among people so that some managers (usually called "line") concentrate on the more dominant objectives and other managers (usually called "staff") think about the deferrable. Such distinction is better than nothing because a less-pressured staff does spend more time on deferrable activities than line managers might. Nonetheless, integrating deferrable objectives into each managerial position is more desirable because it facilitates their actual implementation. Such integration of more and less deferrable objectives dissolves the largely artificial and obsolete distinction between staff and line, counteracts staff's inclination to be unrealistic and overly elaborate, and opposes line's tendency to be consumed by response to present pressures.

Concentrating on one thing at a time is good advice for most people and managers; it conveys order and simplicity. Nonetheless, the behavior of very exceptional managers and professionals contradicts this theme. They tend to be people who can maintain several major and minor projects simultaneously. Their key is ability to shift from project to project fairly frequently. Their objective is to progress on all.

What is involved here is personal flexibility of course. But more fundamental is a sense of the difference between *calendar* or *elapsed* time and *quantity* or *expended* time. Emphasis on single projects is often dictated by deadlines. To meet the due date, time is expended as necessary. Obviously, such a pattern is sometimes essential, but in general it is less desirable than expending time early and in intermittent chunks on a number of different activities maintaining momentum on all. Such a pattern reduces time expenditure because empty waiting and delay time that inevitably crops up on one project can be utilized on another. Under crash project conditions emphasizing calendar time, dead time tends to be wasted. The most effective time managers shift from activity to activity thus using the quantity of time most efficiently.

As a general principle and in the long run, concern for the quantity of time is more desirable than emphasis on elapsed time. Unfortunately, the calendar tends to dominate and massive attention is devoted to the activity with the nearest deadline. Much time is thereby expended.

Every person must make a frank judgement about his own flexibility. For those of us who are less flexible (and this is probably most managers), juggling multiple activities often means dropping one or all. Single-minded concentration on just a few makes sense for most of us. In short, discretionary time is a valuable resource that should be husbanded and focused on activities that promote organizational objectives—the more deferrable as well as the dominant.

REFERENCES

1. To assist in this analysis necessary for reducing response time and concentrating discretionary time, Joseph Trickett suggests that each activity be placed along the four continuums below:

A. INTRINSIC IMPORTANCE

VERY IMPORTANT must be done	IMPORTANT should be done	NOT SO IMPORTANT may not be necessary, but may be useful	UNIMPORANT can be elminated entirely

B. URGENCY

VERY URGENT must be done NOW	URGENT should be done soon	NOT URGENT long-range	TIME IS NOT A FACTOR

C. DELEGATION

MUST BE DONE BY ME I am only person who can do it	CAN BE DELEGATED TO (A)	CAN BE DELEGATED TO (B)

D. VISITATIONS AND CONFERENCES

PEOPLE I MUST SEE EACH DAY	PEOPLE TO SEE FREQUENTLY: NOT DAILY	PEOPLE TO SEE REGULARLY: NOT FREQ.	PEOPLE TO SEE ONLY INFREQ.

Joseph M. Trickett, "A More Effective Use of Time," *California Management Review,* Vol. 4 No. 4 (Summer 1962). © 1962 by The Regents of the University of California. Reprinted by permission of The Regents.

2. One of the most often repeated stories about time management is the

one about efficiency expert Ivy Lee and Charles Schwab, the president of Bethlehem Steel many years ago. It goes comething like this. Schwab asked Lee to give him some advice on how to speed up management activity. "I'll gladly listen and pay you anything within reason that you ask."

"Fine," said Lee. "I'll give you something in 20 minutes that will step up your doing at least 50%."

"O.K.," said Schwab. "Let's have it."

Lee handed Schwab a blank piece of paper and said, "Write down the six most important tasks that you have to do tomorrow and number them in order of their importance. Now put this paper in your pocket and the first thing tomorrow morning look at item 1 and start working on it until you finish it. Then item 2 and so on. Do this until quitting time and don't be concerned if you have only finished one or two. You'll be working on the most important ones first anyway. If you can't finish them all by this method, you couldn't have by any other method either; and without some system, you'd probably not even have decided which was the most important."

Lee added, "Try this system every working day. After you've convinced yourself of the value of the system, have your men try it. Try it as long as you wish and then send me a check for what you think it's worth."

A few weeks later Schwab sent a check for $25,000 with a letter saying the lesson was the most profitable he had ever learned.

Whether this story is true or not is unknown, but it is a ridiculous tale with ridiculous advice. Did Schwab really cut off all response to developing problems and issues that others wanted to raise with him? It seems unlikely. What is the unstated criterion for listing in order of importance? Doesn't this system emphasize the short run? And so on.

3. Attorney General John Mitchell said that "The President recognizes that time is one of the greatest assets he has. He is a man who does his homework, and that becomes quite time-consuming." (Quoted in *Time*, June 8, 1970. See also, "How the President Feels the Nation's Pulse," *New York Times*, December 12, 1969.)
4. Drucker, *The Effective Executive*, p. 103; also pp. 24–25.
5. A survey of this research is James M. Utterback's "The Process of Technological Innovation Within the Firm," *Acadamy of Management Journal*, Vol. 14, No. 1 (March 1971), pp. 75–88.
6. Henry Kissinger is quoted from an interview with Associated Press correspondent Saul Pett in the *Rocky Mountain News*, August 23, 1970. In talking about university professors, Robert Merry makes the same point: "The utter impossibility of fulfilling all the invitations whose importance makes them 'musts'—usually a president meets the condition by assigning priorities and trying to do just as many of the higher priority 'musts' as he can." But, Merry goes on to ask: "Is it possible to construct

another alternative? Can the president reserve any time for himself? Can he determine that some of his needs for time for planning, for reading, for investigating are as important as demands from outside?" ("Pressures on the Presidency," p. 4).

7. M. J. Rathbone is quoted in editors of *Nation's Business, Lessons of Leadership,* p. 196. Ray Brown of Duke University writes: "Every administrator should preview his activities each day and allocate his time and interests to those problems and concerns of highest priority." (*Judgement in Administration,* McGraw-Hill Book Co., New York, 1966, p. 160) Yes, but allocate *all* of his time? And what is the criterion for priority?
8. General Eisenhower's conclusion about trivial and important matters is described in Engstrom and Mackenzie, *Managing Your Time,* p. 68.
9. On the electrical industry conspiracy, see Clarence C. Walton and Frederick W. Cleveland, Jr., *Corporations on Trial: The Electrical Cases* (Wadsworth, Belmont, Cal., 1967).
10. Curtis H. Jones "The Money Value of Time," *Harvard Business Review* (July/August 1968), pp. 94–101.
11. The results objectives are from Peter Drucker, *The Practice of Management* (Harper & Brothers, 1954); the process objectives from Warren Bennis, *Organization Development* (Addison-Wesley, Reading, 1969).
12. The quotations are from Holden, Pederson, and Germane, *Top Management.* (© 1968 by McGraw-Hill, Inc. Used by permission of McGraw-Hill Book Company.)

8 Managing Time—Responding

Concentration is effective, but it is vital to understand that it refers only to discretionary time, not response time. Managers may be highly focused in utilizing discretionary time, in the activities which they initiate and control. Yet, they should be much less focused when responding to others. There they should be receptive to what others choose to raise.

A WARNING: THE NECESSITY FOR RESPONSE

Consider the mental patient who suffers from what psychiatry calls "the God complex."[1] *His* time is valuable in comparison with others, *his* time is correct, *he* alone is justified in being unpunctual, and *his* prediction of future events will alone be fulfilled. Although not so disastrously, this distortion is a danger for hard-charging, egocentric executives. In striving to improve their management of time, they must avoid becoming too self-centered in assuming that allocation of their time is the critical variable in the organization.[2] Just as with some management literature, they may assume the following progression:

better management of personal time → improved personal effectiveness → improved organizational performance

In fact, they may be manipulating themselves and others to suit their own desires. The progression might actually be:

better management of personal time → more satisfying personal behavior → deteriorated organizational performance

For example, one survey of 179 executives states: "The telephone does even more [than meetings] to prevent these top-level executives from making the most effective use of their time . . . 87 percent of the executives spend an hour or more on the telephone each day, and 40 percent spend a minimum of two hours."[3] But, what nonsense! No evidence is offered for why talking on the phone is wasting time. Are no important deals arranged on the phone? Is not critical information tracked down? Are no sincere expressions of gratitude conveyed to a distant associate? Of course they are, and this is not wasted time.

Another artcle maintains that many executives agree they waste time listening.[4] "I get up in the morning and the phone rings and things go downhill from there," complains one company president. Another president says that 70 percent of his time is spent just listening and half that time is wasted "on things you'd avoid if you had any way of appraising them in advance." To imply that time is wasted *just because it is used in an unplanned way* is a distortion of values, a misguided overconcern for time at the expense of the job. Back of this distortion seems to be an unspoken assumption that unless a manager controls his own behavior and does only what he intends to do, he is wasting time. Even in profit-seeking business, this is an exaggeration bordering on absurdity.

When responding, managers must consider several problems simultaneously (or very closely sequenced) even though their own initiation concentrates on one issue at a time. This is a central difference between response time and discretionary time. The former is diffuse and problems are simultaneous; the latter is focused and problems are sequential. Short of total isolation from others, it is simply impossible for managers always to do first things first and one at a time.

However attractive autonomy and discretionary control is, a warning is necessary:

(1) No manager can or should manage all his time.
(2) Response time is not necessarily wasted; it can be very effective in contributing to the organization.
(3) Managers should be wary of managing their time in their own personal interest rather than the organization's.

An important problem for executives is keeping themselves informed. Studies suggest that they do not participate in so many decisions and seldom issue orders. Since they are so dependent on information, executives initiate many inquiries and conversations. In spite of the importance of initiating communications, however, high-level executives depend even more upon others bringing information to them. They must be open and receptive to communications from subordinates and associates. That this is an effective style is suggested by our finding that *high-rated general executives spend more time than lower-rated executives advising and discussing with people who come to them.* As a result, higher-rated executives spend less time on communications that they initiate.[5] Apparently effective general executives develop their various relationships so that others feel free to come to them.[6]

DEVELOPING RESPONSIVENESS

Shortly after the death of Winston Churchill, the Lord of Avon (the former Anthony Eden) discussed what he considered Churchill's outstanding leadership characteristic.[7] He said that the indefatigable former Prime Minister always seemed to be available, he was ready to listen, he never cut off a suggestion with a curt dismissal but encouraged elaboration. To be sure, the great man was often formidable and abusive to subordinates who made mistakes in policy implementation. And he could be dictatorial in ordering the conduct of his own policy. Nonetheless, he was invariably receptive to new ideas. It was not a threatening experience to broach a new matter to Churchill. The next day he might assert that the proposal was unacceptable, but the initial presentation was respected. Consequently, he was approached—and some of the unsolicited ideas were good. It appears that similar behavior characterizes effective business executives. Antony Jay terms this approachability "keeping his feedback nerves as raw and sensitive as possible." In *Management and Machiavelli,* he goes on:

> This also involves listening to criticism from his subordinates . . . Churchill was alive to feedback and Goering was not. The successful creative leader is often willing to listen to criticism from anybody from his vice-president to the office cleaner, and his subordinates are often infuriated when they discover the lowly source of the complaints which are being forwarded to them.[8]

Mentally separating response and discretionary time assists in maintaining responsiveness to others. A manager should decide that certain periods are to be primarily response time. He may work on discretionary matters if no one contacts him, but he should not expect to accomplish anything discretionary. If he does, it is an unexpected bonus. This orientation toward response will help him to overcome the natural irritation of having outside events interfere with self-determined plans. Thus, self-discipline is essential to create a climate of openness and responsiveness.

Managers should strive to project an unrushed image; fight to avoid seeming preoccupied with their own concerns; and fight hard to give subordinates their undivided attention. No one expects perfect attention every time, but some is imperative. A subordinate's respect for himself and for his superior will be undermined without it.[9] For example, yes and no do not consume equal time; no takes longer.[10] A manager can say yes in a second or two (presumably after thinking), but saying no should require more time. Therefore flexibility in responding is especially essential when replying negatively. Except when speed is obviously critical, a manager should take the time to explain negative responses. It shows that he valued the other person's proposal. Such behavior uses more time in the short run, but out of more effective relations, programs may be achievable that no amount of time would otherwise make possible.

An additional note: *research suggests that effective executives conduct more of their numerous discussions in intimate two-person talks than in group meetings or via correspondence.*[11] Effective executives seem to throw out invitations for individuals to come see them personally rather than waiting for the next scheduled meeting. Obviously, this is an untidy way to operate. It consumes much precious time. Yet, executives undoubtedly derive benefits from a two-person style: it facilitates the frank expression of opinions and ideas that they so desire, with less danger of antagonizing others that might be present in a group meeting.

Availability and openness to subordinates and associates recognizes that managers must be responsive or they will lose touch with their organizations. They cannot possibly initiate enough interrogations to maintain control. Being there on the scene and available is a necessity because timing is so important. If the executive is not available when needed to supply information or support, his subordinates may lose interest and drive.

By controlling all their time, reducing response time, and lessening others' autonomy, some managers may centralize the organization in order to reduce ambiguity and threat, but at the cost of future performance. They could be guilty of the hazard of efficiency—efficiency that wastes the time of others.

REFERENCES

1. The so-called "God-complex" is described by Ernest Jones in *Essays in Applied Psychoanalysis* (Hogarth, London, 1951). Another psychiatrist writes: "We all know the individual who cherishes and hoards every moment, but there are also people who not only enjoy wasting their own time, but also take a special delight in dissipating other people's time into the bargain. Subjective time for such people is symbolic in the sense that the value placed upon it is displaced from some source of which they are unaware. Some theorists of the psychoanalytic school have suggested that hoarding time is an anal characteristic just as the dissipation of time signifies a promiscuous scattering of seed." (Cohen, "Subjective Time," in Fraser, *The Voices of Time,* p. 272).
2. Leo B. Moore writes: "By their efforts to solve their own time problems, many managers only compound their organizations' time problems. They fail to realize the extent to which their poor habits affect people around them. They do not schedule their time effectively and accept the resulting excessive demands on their time as necessary condition of their employment as managers." (Managerial Time," p. 83).
3. Quotation from "How 179 Chief Executives Waste Their Time," *Business Management* (March 1968), pp. 12–14. The survey was conducted by by Daniel D. Howard Associates, Inc., Chicago.
4. "Time at the Top," *Wall Street Journal,* August 13, 1968.
5. A summary of the data on general executives is as follows:
   - executives rated more effective spend:

16.0 hr/wk advising and discussing with others who have called or come to them (mean for $n = 3$).

6.5 hr/wk consulting and discussing when they have initiated the conversation.

- executives rated less effective spend:

9.5 hr/wk advising and discussing with others who have called or come to them (mean for $n = 6$).

11.3 hr/wk consulting and discussing when they have initiated the conversation.

Difference between 16.0 and 9.5 is significant at .05 level

Difference between 6.5 and 11.3 is significant at .10 level.

6. In support of this openness and responsiveness by the executive a managerial psychiatrist states that "Every successful executive I know invites engagement with himself." (Levinson, *Executive Stress,* p. 130). Of course, this openness cuts both ways; subordinates have responsibilities upward. Levinson also writes: "The subordinate can support by touching bases frequently with his superior. This may be only a casual hello daily or taking extra pains to keep the boss informed. Many things, particularly feedback about negative or irritating matters, can be easily communicated if there are frequent occasions for doing so. If not, they tend to be stored up for an explosion. This does not mean plaguing the boss—just keeping in touch; do not put him in the position of having to look for you." (p. 123)

A manager comments on a superior whom he admired very much: "I worked for Dick for nine years and he never once sent for me. He always came to my office. This was his way of treating me as an individual with status of my own. I treated my staff the same way. I never sent for them. I would walk into a man's office and talk to him on his own territory." (Quoted by Chester Burger, in *Executives Under Fire,* The Macmillan Co., New York, 1966, p. 95; copyright © 1966 by Chester Burger.)

7. Lord Avon was interviewed on CBS television the day of Churchill's death, January 24, 1965.

8. Antony Jay, *Management and Machiavelli* (Holt Rinehart & Winston, New York, 1968), p. 136.

9. For discussion of the need to give undivided attention to subordinates, see M. M. Feinberg, *Effective Psychology for Managers* (Prentice Hall, Englewood Cliffs, 1966).

10. The observation that saying no requires more time than saying yes is made by Merry, "Pressures on the Presidency," p. 11.

11. A summary of the data on general executives is as follows:

- executives rated more effective spend:

21.2 hr/wk in two-person oral communication (mean for $n = 3$).

12.1 hr/wk in group meetings.

- executives rated less effective spend:

12.6 hr/wk in two-person oral communication (mean for $n = 6$).

19.9 hr/wk in group meetings.

Chris Argyris quotes a manager returning from a sensitivity training session, "I am frustrated because I now realize the importance of face-to-face communications. I'm so far from the general manager that it is not so hot. Has anyone tried to write memos that really get feelings brought out?" (*Interpersonal Competence and Organization Effectiveness*, p. 234).

9 Time and Decision-Making

Hourglass, scythe, moving arrow, a wife in the guise of a grandfather clock, terrifying Chronos devouring his children—all are part of the rich imagery that man has constructed for time. These images, with their emphasis on consumption, disappearance, and decay, have been the dominant symbols in our exploration of time and management. But, there are also older and more positive figures that stress the creative opportunity in time. The ancient Persion Aron was a winged creature with head and claws of a lion wound around a snake and holding a key in each hand. The keys implied the eternal and inexhaustible creativeness of time. Even more dramatic was the Greek Kairos, a decisive choice point in human affairs represented as a man with wings on shoulders and heels, with scales precariously balanced on a knife's edge, together wth a wheel of fortune and a forelock by which he could be seized.

These images convey the concept of man as decision maker. No idea is more popular today in describing the manager. For some scholars of management, decision-making is *the* distinctive managerial activity. Critical and creative decisions are man's way of interrupting, however slightly, the continuity of time's arrow. Management's ultimate objective with respect to time is not to manage it more efficiently, but to use it for creative growth. In this chapter we shall consider time-related and time-affected aspects of managerial decision-making.

RATIONAL DECISION-MAKING

The conditions for ideally rational decision-making may be summarized as follows: (1) An individual is confronted with a number of different, specified alternative courses of action; (2) to each of these alternatives is attached a set of consequences; and (3) the individual possesses criteria that permit him to rank all sets of consequences according to preference so that he may select the alternative that has the preferred consequences.[1]

Given these conditions, the process of decision-making includes:

(1) Diagnosis—identify and clarify the problem's nature and causes. Give the requirements for a satisfactory solution and indicate limits within which the solution must function.
(2) Find alternative solutions—alternatives will range from doing nothing to finding a way around the difficulty, removing the difficulty, or evn modifying the objective.
(3) Analyze and compare alternatives—compare alternatives as to advantages and disadvantages. Ensure that the alternatives under which a choice is to be made are really the ones that should be under consideration.

LIMITS UPON RATIONAL DECISIONS

Many factors interfere with being perfectly rational in decision-making. Perceptual distortions, psychological biases, fear, and anxiety all play roles. No factor, however, is more central in forcing departure from the ideal model than time. Time is money goes the old aphorism that encourages managers to bypass perfect rationality in order to get on with the task. This is not necessarily bad; indeed it may be perfectly rational when the problem fails to justify the effort. Nonetheless, managers should know how time affects decision-making in order to compensate for the distortions.

Awareness of Problems and Alternatives

The rational model of decision-making assumes that the decision maker is: (1) aware of the problem, (2) aware that he must make a decision, (3) has a set of alternatives, and (4) possesses a criterion for making the decision. But these are big assumptions. Problems seldom pop up with identifying flags. Frequently, managers just do not know that they have problems; these are not "felt." If managers are pressed for time and basically operating satisfactorily, they simply fail to use present time to search for future problems.[2] Few systematically search for or consider alternatives unless the present course is creating stress. But creative problem-solving must be preceded by continuous saturation and search. Furthermore, search is essential to avoid crisis decision-making.

Some of the common conflict between line managers and staff specialists derives from this impact of time on search. Operating managers feel time's pressure. As we have seen, their job demands are explicit. They tend to seek predictability and equilibrium in their technical processes and interpersonal relations. In contrast, staff wants to do something different, in effect to upset the equilibrium. Since staff may not be consulted enough by line managers to suit the specialists, staff sometimes has excess time on its members' hands. They begin to initiate, to look for problems that line is not aware of, to propose that line managers give up present predictability, consider problems that only the specialists sense, and make decisions that the managers consider unnecessary.[3]

The ideal model of rational decision-making also assumes that the decision maker is aware of all possible alternatives and that he decides after examining all of them. Obviously this is impossible because time is limited. Managers cannot determine all or even most possibilities So they draw up a partial list based on their own experience, that of associates, articles, and advertisements, and hopefully some creative thought. Frequently, managers fail to postpone evaluation until all alternatives are listed; they evaluate as they develop alternatives. The search for alternatives may cease as soon as a satisfactory one is found, even though all possibilities have not been exhausted and a better solution might remain unexamined. Of necessity, time limits search and leads managers to seek satisfactory rather than optimal answers. All of this makes sense, but it limits rationality.

Pressure to Act

Many managers create their own time pressure because of their already described predisposition to action. Achievement-oriented managers with prestigious offices and high salaries tend to assume that in any situation there is something they can do. Hence executives try not to evade action, but to "do something." This pressure to act pushes managers toward selecting an alternative. Of course, the alternative could be to do nothing, but the mere investment of time in the decision-making process creates a movement for using that investment—and use is more apparent in deciding on change. In other words, time may oppose initiation of search, but once initiated, time pressures for a choice of change. The Bay of Pigs fiasco in the early days of the Kennedy administration illustrates this phenomenon: the decision-making process had a momentum that eliminated the possibility of doing nothing and impelled choice from among the poor action alternatives.[4]

Many managers frequently short-cut the decision-making process by combining causal determination with problem definition. For example, in a case discussed with numerous executive groups, a vice-president has constructed a new assembly line to which he has assigned a rather passive and reluctant engineer as foreman. The new unit is not producing at the desired rate. When considering this problem, time and again managers start with the conclusion that the foreman's weakness is the vice-president's problem. This may or may not be so, but such premature judgment dangerously distorts further thought about the matter. The foreman is not *the problem;* the problem is a growing backlog of orders and customer cancellations stemming from lagging production. The weak foreman may be the cause of inadequate production, but he is not the problem. This is an important distinction. Defining the foreman as the problem tends to focus subsequent thought on how to change or dismiss him. Seeing him merely as a cause will at least allow thinking to explore what other causative factors may be involved (such as faulty selection of operators, inadequate training, and equipment bugs).

A further illustration of this distortion can be drawn once again from the Cuban invasion. In the deliberations about Cuba, government officials defined the problem as Castro. Therefore, solutions were oriented around how to get rid of him—since he seemed immune to our exhortations to change his ways. But, Fidel was not the problem. He was a cause, but only one of several having historical roots well before he and his beard appeared. The problem was the threat of Cuba's geography and personnel. How might the land be used against American interests (a point that became all too evident later in the missile crisis) and how could Cuban agents foment terror and revolution in Latin America (as Che Guevara attempted to). These were the problems, not Castro. If Cuba could have been isolated from the rest of the world, its land denied to others, its agents confined, Castro's presence would have been irrevelant to the United States. The proper solution is unknown, of course, but defining Fidel as *the* problem was misleading and dangerous.[5] Time pressure encourages decision makers to confuse problems, causes, and solutions. They need to fight it.

Unfortunately, maintenance of a certain distance from events and problems at critical decision points will be of increasing difficulty for the manager. Modern information systems can supply him with more data than he can possibly use. With such an inflow, a manager can delude himself that he really knows what is going on, that he possesses a greater awareness, knowledge, and sensitivity to local conditions and issues than really exist. He is tempted to respond immediately and unreflectively just because the technology seemingly allows him to. In addition, because multiple printouts can be widely distributed, the executive may lose some opportunity to sit on data about which he feels uncertain. He can less afford the luxury of not deciding because more people know of the situation; their silent expectations exert pressure toward action.

Desire for Speed

In the broad sense, managers tend to place action ahead of diagnosis and to reward speed. This tendency may even be built into our educational system. "Americans tend to correct children rather impatiently," Edward Hall states. "With us, learning is supposed to be endowed with a certain amount of pressure so that the person who learns fast is valued over the one who learns slowly. Some cultures seem to place less emphasis on speed and perhaps a little more on learning correctly. On the other hand, the current educational mode in the United States is to tell the child to guess if he doesn't know the meaning of a word. Not very good training for future scientists."[6] In response to that battalion of troubles that threatened Macbeth in the opening paragraph of this book, Shakespeare's tragic hero exclaims that if an unpleasant decision is to be made, it should be done quickly: "If it were done when 'tis done, then 'twere well it were done quickly."

Decision makers may move more slowly as they gain in experience, wisdom, and assurance. Recently some high executives were requested to write down the name of the contemporary figure that they most admired (alive or deceased, but contemporary with their lifetime). Then they were asked to write a few words indicating why. The two men mentioned most frequently were Dwight Eisenhower and Harry Truman. The reasons given were similar for both: courage and decisiveness. Some people have criticized President Eisenhower as not being decisive, but few question President Truman in this regard. If anything, he made decisions too easily without sufficient diagnosis, but he never shrank from a decision and he rarely worried about past ones. He said that he always went to sleep shortly after his head touched his pillow. Obviously, these executives admired this ability. It was an attribute they wished they had in their own jobs. In his memoirs, however, Truman's Secretary of State Dean Acheson observed that Truman gradually slowed down his decision-making, overcame his tendency to jump immediately into action, and expanded the period of diagnosis. Yet he still retained the ability to reach a decision by a deadline. Acheson advanced some sarcastic observations on a tendency of later Presidents not to decide, but to struggle "to keep their options open." For Acheson, this is merely indecision.[7]

More systematic research with scientists indicates that more creative men work slowly and cautiously while they are analyzing the problem and gathering basic data. Once they obtain data and approach synthesis and decision, they work rapidly. Less creative men spend less time analyzing the problem but more time in attempting to synthesize their material.[8]

Acheson and Truman were primary combatants in the disagreement with General of the Army Douglas MacArthur concerning his conduct of the Korean War. As might be expected, Acheson was extremely critical of MacArthur's violation of Presidential orders not to take American troops to the border of China. The former Secretary of State suggested that the General made this and other decisions without thought for the dire consequences, and that by surrounding himself with admiring yes-men. he insured that no one raised objections. Whether or not this is true of MacArthur is open to question, but it is certainly true that certain characteristics of many managers promote similar dangers of inadequate diagnosis. Achievement and competitive drives can lead to unwarranted acceptance of one's own work and premature decisions.

NECESSITY FOR TRIVIAL DECISIONS

The total range of decisions covers all managerial and organizational processes, so a complete discussion of decision-making is beyond our intention. Almost all advice to managers, however, maintains that they should slough off trivial matters as they climb the organizational ladder. It is suggested that trivial decisions reflect management's absence of vision. Don't deal with details they are told.

Now this advice makes some sense, but also some nonsense. Obviously, if executives allow their time to be exhausted by details they will lose direction and be dominated by the present at the expense of the future. We shall discuss this at greater length in Chapters 11 and 12. Nonetheless, the superficial importance of a decision is not always indicative of whether or not executives should become involved. Small matters that will confront subordinates with great frequency over a long period of time can be large symbols—and on such matters management should be involved.

Small and detailed decisions that affect organizational behavior and attitudes over a lengthy period are legitimate concerns for executives.[9] Consider the appearance of prospective and new employees. Given the views of many youth on individual freedom in hair and dress, high executives should examine their organizations' formal or informal dress codes. A manager in a large electronics firm recently indicated that this was the number one concern of employees (or at least the most mentioned) in his plant—and this when they were laying off people for lack of work. He recommends that executives set an example by accepting new styles in dress. Regardless of whether other managers agree with this advice or not, they still should be involved, because decisions made will affect many people over long periods of time, thus taking on symbolic content with respect to the company's concern for employees and openness to new ideas. Symbolic, long-lasting decisions reflect a critical leadership dimension.

The timing of a decision can also be symbolic. Some recommend that a newly appointed executive take action quickly—perhaps within the first twenty-four hours. This action may well be a detailed symbolic action (like abolishing preferential parking spaces for managers or eliminating that old relic of nineteenth-century capitalism, payment by the hour). Such a decision is intended to dramatically and quickly set a tone for the new administration. The history of American Presidencies suggests that during the first period in office, a manager has the most freedom to make innovative decisions. An action at this point will be remembered much longer than later decisions and, if he neglects to act early, people will continue in old habits.

The inverse is also true. The absence of early actions creates an impression that colors later perception of the executive. Several political holdovers from the Johnson era and even career civil servants commented about this aspect of the Nixon administration. One old-timer dating back to Franklin Roosevelt's day thought the Nixon administration was *the* slowest in even getting its own men into office. Apparently, the new White House team was not prepared or did not care to make a substantial effort to gain control of the government in 1969. Even in 1970, vacancies open since inauguration were reported to exist. Of course, this inaction might have been symbolic of a larger decision having national significance: that is, the administration's original intention to reduce the activism of federal government—a bright or dull symbol depending on one's political perspective, but one that the Nixon administration could not stick to anyway.

REFERENCES

1. For a comprehensive discussion of managerial decision-making, see David W. Miller and Martin K. Starr, *Executive Decision and Operations Research* (Prentice-Hall, Englewood Cliffs, 1960). See also Richard W. Pollay, "The Structure of Executive Decisions and Decision Times," *Administrative Science Quarterly,* Vol. 15, No. 4 (December 1970). Pollay indicates that decision making takes longer to choose from four alternatives when two are easily rejected than when all four are apparently equal.
2. On the tendency to neglect the important search activity that is the first phase of decision-making, see James G. March and Herbert A. Simon, *Organizations* (John Wiley & Sons, New York, 1960).
3. Elsewhere, I have described at length the staff-line conflict growing out of different time pressures in industrial engineering and manufacturing. See Ross A. Webber, "Innovation and Conflict in Industrial Engineering," *Journal of Industrial Engineering,* Vol. 18, No. 5 (May 1967), pp. 306–313.
4. For details on the Bay of Pigs decision, see Theodore Sorenson, *Kennedy* (Harper & Row, New York 1965) and Arthur M. Schlesinger, Jr., *A Thousand Days* (Houghton Mifflin Co., Boston, 1965). Writing about a later and more effectively managed crisis (the Cuban missile confrontation), Robert Kennedy said: "I often thought afterward of some of the things we learned from this confrontation (on Soviet rockets in Cuba). The time

that was available to the President and his advisers to work secretly, quietly, privately, developing a course of action and recommendations for the President, was essential. If our deliberations had been publicized, if we had had to make a decision in twenty-four hours, I believe the course that we ultimately would have taken would have been quite different and filled with far greater risks. The fact that we were able to talk, debate, argue, disagree, and then debate some more was essential in choosing the ultimate course. Such time is not always present, although, perhaps surprisingly, on most occasions of great crisis it is; but when it is, it should be utilized." (Robert F. Kennedy, *Thirteen Days: A Memoir of the Cuban Missile Crisis,* W. W. Norton & Co., New York, 1969, p. 111. Copyright © 1969 by W. W. Norton & Company, Inc.; copyright © 1968 by McCall Corporation. Reprinted by permission of the publisher.)

5. The concentration on Castro as the problem in Cuba is especially evident in the oral-history transcripts in the John F. Kennedy Memorial Library. Reports on conversations between Senator Smathers of Florida and President Kennedy indicate consideration of the implications of assassinating Castro as a solution. (Henry Raymont, "Oral-history Fills Out Record of JFK Years," *Denver Post,* August 23, 1970).

6. On time pressures in learning, see Edward T. Hall, *The Silent Language* p. 55. The quoted passage is from *The Silent Language,* by Edward T. Hall. (Copyright © 1959 by Edward T. Hall. Reprinted by permission of Doubleday and Company, Inc.)

7. For Truman on himself, see Harry S Truman, *Year of Decisions,* Vol. 1, *Memoirs* (Doubleday & Co., Garden City, 1955; and *Years of Trial and Hope,* Vol. II, *Memoirs* (Doubleday & Co., Garden City, 1956). Of President Truman, Dean Acheson writes: "His judgement developed with the exercise of it. At first, he was inclined to be hasty as though pushed out by the pressure of responsibility, and—perhaps also—by concern that deliberateness might seem to be indecisiveness. But he learned fast and soon would ask, 'How long have we got to work this out?' He would take what time was available for study and then decide. . . . No one can decide and act who is beset by second thoughts, self-doubt, and that most enfeebling of emotions, regret. With the President, a decision made was done with and he went on to another. He learned from his mistakes (although he seldom admitted them), and did not waste time bemoaning them. . . . Not in any sense self-deprecating, his approach was sturdy and confident, but without any trace of pretentiousness. He held his own ideas in abeyance until he had heard and weighed the ideas of others, alert and eager to gain additional knowledge and new insights. He was not afraid of the competition of others' ideas; he welcomed it. Free of the greatest vice in a leader, his ego never came between him and his job." Dean Acheson, *Present at the Creation: My Years in the State Department.* (W. W. Norton & Co., New York, 1969, pp. 731–733. Copyright © 1969 by Dean Acheson. Reprinted by permission of W. W. Norton & Company, Inc.)

8. On creativity and problem-solving, Berelson and Steiner write: "In a problem-solving situation, the more creative man works slowly and cautiously while he is analyzing his problem and gathering his data. Once he obtains the basic data and approaches the point of synthesis, he works rapidly. The less creative man spends less time in analyzing the problem but more time in attempting to synthesize his material." (Bernard Berelson and Gary A. Steiner, *Human Behavior,* Harcourt, Brace & World, New York, 1964, p. 227. See also Morris I. Stein and Shirley J. Heinze, *Creativity and the Individual,* The Free Press, New York, 1960).

9. Townsend points out that nothing is below the dignity of a high executive, no detail is too low. He writes: "A good manager is a blocking back whenever and wherever needed. No job is too menial for him if it helps one of his players advance toward his objective." (*Up the Organization,* p. 102)

10 Individual and Group Decision-Making

"A camel is a racehorse designed by a committee." So goes one of many aphorisms on the inadequacies of groups as decision-makers. In spite of the fun in such complaint, however, committees continue to make decisions, because they offer advantages in breadth of experience, varied knowledge, absorption of antagonism, and mutual support.

COMPARING INDIVIDUAL AND GROUP PERFORMANCE

To compare some aspects of individual versus group performance, numerous exercises have been conducted with managers and students. In our research, groups take longer than the average of individuals working alone (50 percent longer where individuals can complete the task), but over three-fourths of the groups produce better performance (an average of 30 percent better). Most groups, however, are worse than the best individual in the group.[1]

These findings suggest that a group may be an advantage where being correct or avoiding mistakes is of greater importance than speed. The group does seem to improve on the performance of most people.[2] In short:

(1) The best individuals are usually better than groups as to accuracy, speed, and efficiency.
(2) The average individual is faster and more efficient than most groups, but he makes more errors.
(3) Groups are more accurate but slower than most individuals.

The validity of these conclusions varies somewhat with group composition however. Age and hierarchical level of the group members affect the results. Various groups have participated in our research: among others, they include high-level general executives averaging forty-seven years of age, forty-year-old middle managers, thirty-two-year-old young managers, and graduate and undergraduate business students about twenty-five and twenty years old respectively. The findings are distinctive as to their clarity. Working as individuals, there are no significant differences in task performance among the different categories of people. *Yet. lower-level young men are more effective in utilizing group decision-making than older, higher-level managers.* The difference between group performance and average individual performance decreases with increasing age and level of the group members. Younger groups improve more on individual performance. In fact, college students constitute *the only groups* that have higher group scores than their best individual. Among all others, the best individual is better than the group.[3]

So younger people seem to be more effective in utilizing groups for decision-making than are older and higher-level managers. Time and age seems to weaken ability to work jointly with others—or perhaps the younger students have come through an education system that has placed greater emphasis on group activity. The reasons why are beyond the scope of this book, but they might also include: less sensitivity to status, more personal flexibility, greater willingness to express opinions, and more "team spirit." Nonetheless, all groups offer advantages and disadvantages with respect to most individuals—more correct answers, fewer errors, but slower.[4]

Whether the advantages of group decision-making justify the additional time depends upon three critical factors: (1) whether speed is essential (as when a military unit is under attack or a prospective customer is threatening to terminate negotiations), (2) whether an incorrect decision can be tolerated (as it cannot when making important defense decisions in Washington or styling decisions in Detroit), and (3) whether the organization has an exceptional individual who would be hindered by a group (as the United States had with Mr. Lincoln in the White House or as a few corporations have had with their entrepreneurial giants).

Time pressures adversely affect individual decisions, but perhaps the impact on group decisions is even greater. Conformity and silence often result. Laboratory research suggests that group solutions are superior only when the initial judgments of the individuals in the group are not identical; that is, a range of possibilities is present. This advantage is lost if the situation is dominated by an authoritarian leader aided by a shortage of time. Listen to one manager's complaint:

> Here we don't respect differences, we clobber them. Let's say we're in a meeting, the agenda is full and time is running out. So the boys begin to get edgy. Finally, the boss says "Okay, it'll be this way and that's it." Here you shut up, say "yes, sure," and even if you're smart, add "that's a good idea."[5]

Thus, time pressure and a hurried agenda help an authoritarian manager to control a meeting. What time pressure does is limit the open expression of conflict. And it does this even in nonhierarchical groups. If all members of a team recognize and accept the time limitation, each individual is much less likely to argue with his fellows. The group may even reward its members for loyalty and teamwork when concensus seems to dominate. Yet, the whole situation may be a sham, an artificiality of a shared desire to get things done in minimum time.

If an organization operates with a continual series of crisis, conflict may be continually suppressed. Does this suppression cost management anything? It probably costs individuals in the psychomedical area, but it is difficult to judge whether the organization is hurt. Suppression of conflict in the short-run facilitates task performance in the short-run, but risks undermining the organization's capacity to innovate and adapt to changes in environment and mission. A manager may be able to lessen open conflict among his staff by exerting time pressures, but he can probably keep it up for only a limited period. Permanent suppression of conflict leads to organizational self-delusion and eventual atrophy.

TIME ALONE AND TIME WITH OTHERS

It is not really an either-or proposition between group and individual decision-making. It depends upon the stage within the decision process.

Diagnosis and Choice

A clue to how this works is offered by the following incident: A group of young women were working on an exercise problem which went well until they reached the solution stage. At this point, one of the girls (an extremely bright Phi Beta Kappa graduate) jumped up exclaiming, "I can't stand this anymore!" With that, she bolted from the room. To everyone's surprise, she returned quickly with a correct solution. She explained that she left when she had enough information; the group's questions and talk was a hindrance after this. The noise and confusion interfered with her thought—so she left.

What happens is that the group is helpful in diagnosis—analyzing the problem, the objective, and the limits upon the solution. This is apparent in the pattern of questions: they are asked in batches, with one person's triggering off three or four from others. Then a silence follows, then another cluster of questions, and so on. Everyone feels free to utilize the information developed by others. So the group aids in collecting information necessary to solve the problem, but tends to interfere with the detailed formulation of a solution. In short, a group may be helpful in the analytical stage of decision-making (perhaps three-fourths of the total time to make the decision), but a hindrance to final evaluation and selection.

During the decision-making and creative process, therefore, some time should be spent with others and some time alone. The entire operation is not suited to either state exclusively. The thinker as well as artist must isolate himself from the everyday world or ideas and images will never come to mind.[6] But this detachment requires prior reality to stimulate and channel it. Even for the artist, although there is a time when he needs to be left alone without interruption in order to produce his work of art, there is a time before that when he very much needs other people for ideas, facts, guidance, and practical suggestions. Other people are especially helpful for analyzing problems. After this, it may be appropriate to withdraw to a secluded wood or analogous location for cerebral cogitation. For example, Presidents Johnson and Nixon illustrate two different styles of decision-making. Johnson reportedly preferred to operate in a group with much verbal discussion (much of it one-way from him) with the decision emerging directly from the talk. No sharp transitions from diagnosis to selection were apparent. In contrast, President Nixon desired time alone in order to evaluate and choose by himeslf free from the distractions of other people. Here is how Roger Blough of United States Steel considered this aspect of decision-making:

> Once you think you have learned something about these factors (in the decision), I think it's most useful to kick the idea around among your associates—see what their views are. After you've done that, I think it's important to make sure you are not so close to the problem that one factor is given more weight than it deserves. One way to make sure this doesn't happen is to sleep on it overnight at least and maybe a week or two depending on the nature of the problem.[7]

For another example at a much lower level, an exceptionally effective young industrial manager consciously utilizes his alone time. After important meetings his next step is to go hide somewhere and think. He tries to imagine himself as the others in the situation: "Well, I am someone else on this job," and to answer the question: "What is my interest in it?" He finds it especially valuable when he is extremely involved in a problem and all keyed up.[8]

Time to Think

An additional problem for many Americans (and of course some others too) is that we make a distinction between whether or not we are engaged in *activity*. Just sitting alone trying to capture a sense of self is not considered to be *doing* anything (except perhaps in church).[9] The sunbathing assiduously pursued by so many may have little to do with changing their color, but the tanning justifies lying motionless sleeping or dreaming during the day. The sunbather is *doing* something, while just rocking in a hammock under a shady tree is the cartoonist's stereotype for wasting time. In many cultures, simply sitting is doing something; no distinction is made between being active or inactive. This is not a recommendation for indolence, but a suggestion concerning the efficacy of thought. With some oversimplification, but much truth, Bernard Baruch once stated: "Whatever failures I have known, whatever errors I have committed, whatever foolishness I have witnessed in private and public life have been the consequence of action without thought." Theodore Levitt at Harvard Business School advances the same idea in terms more critical of management:

> The management of a large organization is far more difficult than most men who manage are themselves aware. Few in fact are managers. They are custodians. Somehow the enterprise keeps going at an acceptable rate, but usually only because its competitors are no better. Most managers unfortunately do very little thinking in the course of their work. They are entirely unaware of how much they manage by formula, by dogma, by principles, textbook maxims, and resounding clichés; and of how much, in the process, they are forfeiting the one distinction by which we tell man from animal. Their resort to easy generalization is explained by the heavy burdens of their job. But to explain it is not to condone it.[10]

For most executives, obtaining time alone for decision-making is difficult. It tends to be broken into small pieces because telephone and personal interruptions reduce unbroken periods to just minutes. Managers complain loudly about these interruptions, but modern middle-class life may accept and require such activity. Interruptions feed the desire to be important and indispensable. And our life style interferes with thinking alone. Ubiquitous portable radios, paperbacks, and personal television sets have habituated Americans to never being really alone.

Specific positions entail their particular difficulties obtaining time alone for thought and decision-making. As was mentioned in Chapter 6, at a conference of church officials it was suggested that the culture of the institution mitigates against withdrawal by the bishop/manager. Responsiveness and openness to staff is ingrained in the expectations of all parties as part of their organizational and religious mission. Much the same point was made by a dean of a graduate school of social work. To her, social agency administration is a social work process, not organizational management. For some church and social work officials, to absent themselves from the office, to postpone a meeting, or to refuse to see anyone who comes to see them is almost sinful. Consequently, these managers control little of their decision-making time.

During a management development program for a state government, the participants were discussing the message of this and the preceding chapter—the need to expand diagnosis and to isolate oneself in order to think. One participant, the warden of the state penitentiary complained, "Yes, but everytime I close my office door to sit and think, some wise guy stands outside my window and yells 'There they go!'"

REFERENCES

1. The following table summarizes my preliminary findings:

Comparison of Individual and Group Performance

	Individual $n_1 = 240$	*Group* $n_2 = 48$
Mean time to complete vocabulary test	4.5 min	6.8 min
Mean score (number correct out of 25 questions)	13.2	17.5
Mean errors (number incorrect)	11.8	7.5
Mean efficiency—number right per minute	3.0	2.6

2. Some other findings on group behavior as summarized by Berelson and Steiner, *Human Behavior,* p. 355: Under the following conditions, the solution to a given problem by a small group, after discussion, is more satisfactory than the average or composite solution of the individuals making up the group:

- When the problem has a definite and identifiable solution.
- When the initial judgments of the individuals in the group are not homogeneous, so that a range of possible solutions is initially available to the group for its consideration.
- When the task requires that each member make a judgment about the same matter.
- When rewards and punishments are given to the group as a whole rather than to individuals within the group.
- When the information or skills needed for the solution are additive.
- When the task can be subdivided.
- When the task includes "traps" that single individuals might miss.

Michael Argyle concludes: "The collective judgment of a group is superior to the judgment of most of the individuals. Two distinct processes are involved—firstly, discussion leads to the improvement of individual judgments, and secondly, the combination of individual judgments is advantageous." (*The Scientific Study of Social Behavior,* Methuen, London, 1957, p. 119) Another review of the literature states: "In general, in the evaluation of the relative quality of the products produced by groups in contrast to those produced by individuals, the group is superior. The superiority of the group, however, all too frequently, is not as great as would be expected from an interactional theory. In many studies, the product of the 'best' individual is superior to that of the 'best' group." [Irving Lorge, David Fox, Joel Davitz, and Martin Brenner, "A Survey of Studies Contrasting the Quality of Group Performance and Individual Performance 1920–1957," *Psychological Bulletin,* Vol. 55 (1958), p. 369. See also Norman R. F. Maier, "Assets and Liabilities in Group Problem Solving: the Need for an Integrative Function," *Psychological Review* Vol. 74 (1967).

3. The following table summarizes my preliminary findings on how age/position level of group members is related to performance.

How Age/Position Level Is Related to Individual and Group Performance

Subjects	*Mean score (no. correct)*	*Mean time (min)*	*Mean group effectiveness (group score minus mean of indivs. in group)*	*Mean group excellence (group score minus best indiv. score in group)*
Executives				
mean age 47				
40 indiv.	13.4	4.0		
8 groups	15.8	9.0	+1.8	−1.4
Middle managers				
mean age 40				
55 indiv.	13.3	4.2		
11 groups	15.1	6.2	+1.7	−2.9
Young managers				
mean age 32				
90 indiv.	13.2	4.2		
18 groups	15.3	5.3	+2.3	−3.1
M.B.A. students				
mean age 25				
90 indiv.	14.3	4.7		
18 groups	16.3	6.0	+3.7	0
B.S. students				
mean age 20				
40 indiv.	11.5	5.5		
8 groups	17.5	6.0	+3.5	+1.5

4. On the problem of the time that groups consume to reach decisions, Merry tells how one college president recommended to his fellows that discussion of issues continue until the faculty had approached unanimity. Another president in the audience could no longer contain himself and burst out, "I don't mind dallying with time, but I can't tamper with eternity." ("Pressures on the Presidency," p. 10)

5. The manager is quoted in Argyris, *Interpersonal Competence and Organizational Effectiveness,* p. 69. Argyris goes on to observe that the major theme in his feedback sessions with the company was the scarcity of time: "There never was enough time. Time seemed to work against the desire to make adequate decisions. If time was lacking, then the incompetent subordinate might use it to protect himself. As long as he knew that the leader saw himself as being responsible for establishing the facts, then all the subordinate needed to do was procrastinate long enough and time would coerce the leader to take action. An executive added that when this occurred the subordinate would probably no longer need to feel responsible for making decisions." (p. 108)

6. On the artist's alternating alone time and social contacts, see Edgar Wind, *Art and Anarchy* (Faber & Faber, London, 1963). De Grazia makes a similar point: "The artist as well as the thinker is a child of slow time. He must be able to detach himself from the everyday world else ideas and images will never come into his mind." (*Of Time, Work and Leisure,* p. 421)

7. Roger Blough is quoted in editors of *Nation's Business, Lessons of Leadership,* p. 221. A French sociologist observes that taking this time is difficult: "The man of the technical world [has] effectively suppressed the respite of time indispensable to the rhythm of life; between desire and the satisfaction of desire there is no longer the duration which is necessary to real choice and examination. There is no longer respite for reflecting on choosing or adapting oneself, or for acting or wishing, or pulling oneself together. The rule of life is: no sooner said than done. Life has become a racecourse composed of instantaneous variations of the universe, a succession of objective events which drag us along and lead us astray without anywhere affording us the possibility of standing apart, taking stock and ceasing to act." (Jacques Ellul, *The Technological Society,* Alfred A. Knopf, New York, 1964, p. 329) Blough's remarks would suggest that he tries to fight this.

8. The very successful young manager, Mr. Bates, is described by Dill, *et al.,* in *The New Managers.*

9. In the self-awareness growth activities at Esalen in Big Sur, California, the unfamiliarity of time alone with oneself is put to use. William Schutz writes: "[We] assign each person a 30 minute period during which he must be alone and reflect on the events he has experienced. He is to focus on himself, his feelings, and his rlations to others, . . . this activity provides an opportunity for the participant to explore himself and to provide a framework and support for introspection of a serious nature.

"For many people this is an electrifying experience. Several report that they had never used their time alone in this way before. For many, it affords an opportunity to consolidate experience. Also, it provides a way of using alone time that may well be valuable in the future to many people." (William C. Schultz, *Joy: Expanding Human Awareness*, Grove Press, New York, 1967, p. 62. Copyright © 1967 by William C. Schutz. Reprinted by permission of Grove Press, Inc.)

10. Theodore Levitt, "The New Markets—Think Before You Leap," *Harvard Business Review,* Vol. 47, No. 3 (May/June 1969), p. 67. Kahn and Wiener advance similar views in *The Year 2000,* p. 393: "Most administrators dislike debating or thinking about fundamentals, even when vague, implicit, and half-formulated views obviously are governing choices, and when some searching debate is clearly desirable. Administrators resist even more 'unnecessary' discussions that may become unpleasant or divisive. They tend to resist still more the very basic or very speculative thinking that may be essential to raising issues about the future, in part because of a well-grounded feeling that such thinking and discussing are usually unproductive and expose those who make the attempt to criticism, bureaucratic animosity, or ridicule. Yet it is often necessary to be courageous—or seemingly irresponsible—in suggesting and defending far-fetched issues in argument, if unprecedented but crucial considerations are to be discovered and appreciated. It is necessary to spend some time and energy in a process that frequently leads nowhere for the sake of the instances in which something new is learned." (Macmillan, New York, 1967; Copyright © 1967 by The Hudson Institute, Inc.)

11
Past Time

"All history is bunk," Henry Ford is reputed to have once announced. What Mr. Ford was suggesting is that managers are concerned with the future, not the past. Their thoughts should be about future conditions, future events, and future returns. Accounting and finance teach us that past costs are gone; they should play no role in decisions about the future. In evaluating human performance, the control system should not emphasize punishment for past deeds, but should be oriented toward improved organizational performance in the future. Certainly the major concern of this book is present and future managerial behavior. Nonetheless, the past intrudes on the present and influences the future. The past can be vital for understanding the present and visualizing the future.[1]

Ironically the Ford Motor Company dramatically illustrates the relevance of the past to present decisions. Henry Ford's inadequate understanding of history contributed to the company's decline in the late 1920s. In his memoirs, Alfred Sloan, longtime chief executive at General Motors, points out that Ford did not recognize the historical development of public taste and the desire for increased choice and luxury.[2] Sloan understood and General Motors responded in their product policy. Ford did not, so sales of the Model T declined more precipitously than had horse-drawn buggies a generation before. In 1927 almost no Model T's were sold. Ford had nothing else to offer until a year later when the Model A was unveiled—a car that Sloan thought to be old-fashioned even at its introduction.

TYRANNY OF THE PAST

Blind obedience to the past is characteristic of certain neurotic individuals and their organizational counterparts. Psychiatry tells us that a sense of identity is necessary in order to escape the pull of the past. Patients with compulsive neurosis demonstrate continued obedience to old conditioning demands; they rigidly obey mechanical orders in their historical sequences. They stick to tyrannical inner schedules out of fear of real life. Slavish adherence to the past, to the "old ways that are best" is a means of warding off the unknown with its associated anxiety. In some children the compulsion to repeat pleasure *ad nauseum* and unto utter exhaustion can be observed. Some adult neurotics refuse to recognize passing time and most of us occasionally cling anxiously to the past.

Henry Ford's supposed contempt for the past does contain a healthy aspect then. Managers must not let the past tyrannize the present; they must not become slaves to past glories. A healthy skepticism (but not ignorance) is helpful in questioning the lessons of the past and in reexamining them in the light of the present.[3]

Clinging to the Past

History and habit are related. Historical obedience through habits can be harmful to management. This is true both for habits of thought concerning the organization and behavioral habits relating to others. The military provides some classic illustrations of this adherence to the past in the face of change. Two of the best (or worst):

> In 1897 a British Naval Officer, Sir Percy Scott, designed a modification to naval gun mountings that allowed the weapon to be aimed continuously as the ship rolled with the sea. Prior to this, the pointer had to anticipate the roll of the vessel and fire before the sight was on the target. The practice firings indicated an improvement in accuracy of some 3000%!
>
> In 1900 young Lieutenant William Sims of the United States Navy met Scott while on China patrol. Sims was impressed and

> disturbed that the U.S. lagged so far behind. He wrote to the Navy Department and the Bureau of Ordnance, but with no success. He was unknown and so far away. Besides, hadn't the American Navy just established its power in the Spanish-American war with a crushing defeat of the enemy in Manila Bay?
>
> The United States had won that battle, of course, but under the old system of gunnery; it has been estimated that of some 9,500 shots fired in that engagement, only 121 found their mark. History and success apparently meant too much for Washington officials, however, and Sims was rebuffed.
>
> Success came only when the ambitious Sims put his career on the line and went directly to President Theodore Roosevelt—a man with some respect for the past, but more for the future. The United States implemented the ordnance modification and Sims ended up as senior Naval officer in World War I.
>
> To give the British military their due, there is a tale (of unknown veracity) about their use of a venerable, formerly horse-drawn artillery piece in the early days of World War II. Guns were in short supply and this old piece dating back to the Boer War was being used for coastal defense. It was slow firing, however, and a time study man was brought in to determine how the firing rate might be improved.
>
> The engineer observed a five man crew in action and took some slow-motion pictures. Studying the actions of the crew he noticed something odd: a moment before firing, two members of the crew ceased all movement and came to attention for a three second interval extending through the firing of the weapon.
>
> He consulted an old Colonel of artillery. At first the elderly officer was also puzzled, but he finally exclaimed, "I have it. They are holding the horses." The horses, of course, had long since gone the way of the Boer War.

As Elting Morison sums up the lessons of the tale: "This story . . . suggests nicely the pain with which the human being accommodates himself to changing conditions. The tendency is apparently involuntary and immediate to protect oneself against the shock of change by continuing in the presence of altered situations the familiar habits, however incongruous, of the past."[4]

To the business manager these stories epitomize the limitations of the military mind. But all minds are quite similar in this regard. In the engineering department of a large corporation, a bias for the past betrayed itself in the consideration of alternatives. In presenting proposals for management decisions, engineers listed and investigated alternatives in a habitual order: first was what was being done already; second was a modest change built on the present; third, something someone else in the company was doing; fourth, something some other company was doing (there might be several alternatives at three and four). Only after these was there an alternative involving anything really new. For example, on one project they were transporting raw material in 25-gallon fiber drums. The alternatives were listed as follows: (1) the present drums, (2) smaller cardboard cartons, (3) paper bags similar to ones being used in another department, (4) aluminum bins similar to those used by a local bakery for flour, and (5) deflatable rubber bins that someone had read about in *Factory* magazine. Alternatives 4 and 5 were never really considered. Why not list and consider alternatives in random order? Or reverse order of familiarity?

Even a successful company like General Motors may let the past dominate the present. Chevrolet's near-disastrous difficulties in the 1920s with a "copper-cooled" engine may explain its reluctance to pioneer in anything but the cast-iron block, water-cooled, internal combustion engine. It has been suggested that the 1959 introduction of the Corvair aluminum air-cooled engine was accomplished only by secrecy; lower officials supposedly deceived corporate executives about the specific engine design of the Corvair until styling was virtually completed. The subsequent demise of this model can hardly provide much of a counter to the tradition.

Not Recognizing Change

Consider the owner/manager of a small printing establishment. Concentrating on straight letterpress printing for a wide variety of local and regional customers, he was very successful. When offset printing came along, the owner felt little need to add the new technology because "I don't have room," "I don't have time," "It produces lousy copy," "None of my customers would use it," and so on. This resistance to new technology went along with a backward-looking view toward certain customers. The largest institutional customer was the local government, followed by the local hospital, schools, and churches. They were good customers, even if slow payers, but the owner/manager's perception of them was distorted by a certain contempt for nonprofit institutions. Namely, they were supposed to be badly managed, inefficient, and not greatly concerned about costs—especially politicians who only collected taxes and spent money. But, he was wrong. Cost pressures on these institutions were great and increased with affluence rather than decreased. Quality was a lesser concern (and offset quality improved with time anyway). He lost an opportunity to grow because of his misperception of historical trends.

Some large corporations who grew under the drive of a commanding figure such as Watson or Sarnoff can fall prey to an unconscious sanctification of past policies. They unknowingly endow their heroic leader with infallibility. As a result, his ideas are sometimes applied to changed conditions. In the face of increased competition and rapidly changing dress and social values, Thomas Watson, Jr., in 1971 attempted to resurrect the Arrow-collar, white-shirt, short-hair look of the fifties in order to convey what he thought was an image of vitality. His success or failure is open to question, but such backward-looking action seems inappropriate to an organization with a forwarding-looking technology.

Finally, consider the following event which we have discussed with many managers: in mid-1969 during the national debate on our policy in Indochina, a group of engineers at a large American defense contractor submitted a white paper to management announcing that they would no longer work on a certain product that was being manufactured for military use in Vietnam. What should management do? How should management respond? Most managers react strongly that this is a violation of a very fundamental organizational principle: that management should determine product and customer policy while engineers and other employees should design and manufacture. "Fire them all!" is a not uncommon response to the situation. Getting rid of the offending parties is seen as a method for maintaining managerial control. Perhaps so, perhaps not, but at least a minority of thoughtful managers have asked if the majority is not responding from history rather than reality. Is the inclination to fire the young engineers a defensive reaction to the possibility of change? Does it imply that prevailing managerial philosophy is becoming obsolete? In fact, is not participation of the designers/makers in determining product policy the wave of the future? Lessons from universities and governments suggest that it may well be.

In this last example, fear of new technology does not explain the reaction. Rather, it is preference for the social and authority patterns of the past—and this is infinitely more corrupting and dangerous.

CODIFICATION OF THE PAST

The past is codified in business through standing plans, policies, procedures, and controls. This is necessary, but still dangerous.

Necessity for Policies and Controls

At a number of large universities, students have organized "free universities" open to all interested parties with no tuition or academic credentials required. Specialized courses on contemporary and esoteric subjects (two examples, Vietnam and the Draft, Body Awareness) are offered by volunteer instructors. The founders' intentions were to provide a wider choice of courses, more flexibility, and greater individual attention than was available in the regular schools with their red tape and impersonality. A student founder of the largest of these free universities recently resigned with a bitter comment about the growth of stifling bureaucracy and inhibiting rules even within the new institution.

This free university was born as a protest against formal university structure and values. At its high point, several thousand students were attending classes. It was apparently successful, yet the founder repudiated it. He simply was not aware of the managerial functions necessary in an organization once it grows beyond small size. In this case, these functions resulted from scarce resources; here the initial policies (and red tape) emerged. Classrooms were limited and were not all the same size. Control had to be exerted over scheduling hours and the number of students enrolled for the courses. Records were required; communications were essential, and exclusion of some students was necessary. This exclusion led to an important managerial function, unpopular decisions, because admission criteria are essential if class size is limited. In addition, professors were in short supply. Class hours had to be arranged to fit their availability, which required information about faculty schedules. Finally, professors were to be given freedom to teach whatever they wanted, but the free university also wanted to provide courses that students desired. Such objectives imply substantial coordination and the setting of priorities.

In short, even in this informal, protest-oriented institution, managerial functions exist. Managers are needed and policies and procedures are necessary to assist them in administering the organization. It was unfortunate that the youthful founder was ignorant that such would be required. Certainly, rules must be subordinated to desired ends, and flexibility should be maintained so that red tape does not become an end in itself, but policies and procedures are necessary because they reflect basic organizational processes.

Consider a business example. In a developmental phase, a company may be quite successful through depending for guidance almost entirely on goals with few policies, procedures, or controls. Eventually, however, controls are necessary and they develop. When they develop there is danger of inverting means and ends. The Xerox Company offers an illustration. For the past fifteen years it has been one of the most exciting organizations in American business. The old family-dominated Haloid Company, manufacturing specialized photographic products, transferred itself into the modern Xerox, and jumped from 25 million to approximately a billion dollars per year in revenue.

In the late 1950s the company was chaotic. Offices were located all over the Rochester area, over delicatessens and in abandoned schools. Job descriptions were few, policies were broad, procedures nonexistent, and controls weak. Yet the company was successful. And it was successful because of top management's ability to define direction. The then chairman Sol Linowitz and President Joseph Wilson saw their roles as entailing the formulation of objectives. Wilson devoted much of his time to selling the Xerox Company and its objectives to his own managers, describing the revolutionary and beneficial impact of its information technology on society—and also pointing out how each manager's own interest would be served if the company advanced. Given fantastic expansion, he was aware that it would be impossible to control the entire company from the president's office. In order to take advantage of a great market opportunity and to exploit technological breakthrough, it was essential that Xerox managers at all levels be committed to manufacturing their products and getting them out into the market as quickly as possible. Premature policies, procedures, and controls would have interfered with the spontaneous cooperation and initiative demonstrated by management.

Nonetheless, standing plans and controls did emerge in the Xerox Company, as they do in all organizations.

Using Past Experience

Why go on reinventing the wheel? Why must the company handle every problem as unique? Standing plans and procedures often begin to develop from this awareness. Sam Johnson wonders how Tom Gilhooley has handled the problem of quality control on his gears.

He asks about it. Tom's answer seems reasonable, and rather than searching for new alternatives, Sam adopts Tom's procedure. Or Mike Stratton has had great success in handling rush orders. How does he do it? Some questions will uncover his methods—why not let everyone know about it? So, an information memo is drafted and distributed. The basis for standing plans and procedures, therefore, rests upon an awareness of the stupidity of not using past experience. Why should management repeat the same mistakes over and over again without taking advantage of the accumulated knowledge of people in the organization? The famous words, in another context, of Justice Holmes apply: "Three generations of idiots are enough!" The company decides to implement some policy that trades upon experience and gives guidance to people facing similar problems for the first time.

Initially these policies and procedures are developed from the organization's own past. After a time, however, management may feel it should take advantage of other people's experience instead of restricting themselves to their own company. Experts provide this knowledge. So specialists, advisers, auditors, and controllers are engaged. Once they are on the payroll, there is a natural tendency for management to see that they are used. Frequently they are granted some authority to impose their experience on the rest of the organization.

This development of job descriptions, policies, and procedures aids efficiency by promoting coordination and predictability. However fine the initiative of Xerox managers had been in the expansion stage, it was stressful and unstable. For a while, people will accept instability, but order, regularity, and predictability become essential in the long run—especially if the growth rate and promotional opportunities begin to decline. Nor do men necessarily dislike the advent of standing plans and controls. Order will be brought out of chaos, probablity out of uncertainty, and simple relief from repeating the same mistakes. Indeed, most managers perceive the development of control instruments as progress.

DANGER IN POLICIES AND CONTROLS

However inevitable and necessary the codification of the past in policies and controls, also inevitable is the development of problems with those plans and controls. One of the so-called management principles of long repute has been the rule that delegated duties should be explicit and specific—with no gaps or overlaps. If this is true, if policies are inclusive and if control measurements are feasible, theoretically each manager need only follow directions. No initiative would be required. Of course, this is an impossible condition. Such perfect delegation, planning, and controlling is not attainable. Therefore, spontaneity is essential. To a greater or lesser degree, every manager must fill in the gaps, work out the conflicts resulting from overlaps, and exercise discretion in following standing plans.

Loss of Direction

If policies are followed blindly, the organization loses direction. There is an inversion of means and ends. To some people, rules, plans, and controls become ends to be followed without thought as to whether they contribute to the organization's objectives. Indeed, such inversion of means and ends is almost synonymous with bureaucracy and organization.[5] Business managers frequently assume that such distortion characterizes nonprofit institutions such as governments, schools, and hospitals. But, it also occurs in business. An advertisement in a prominent management journal once showed a hand holding a notebook embossed in gold letters with the words *Policy Manual.* The notebook was suspended over a wire basket incinerator in which several similar notebooks were burning. Who was the advertiser? A management consulting firm. Their message? That one should not allow standing plans, policies, and procedures to exist unchanged for too long because they get out of date and hinder the organization instead of helping. Better to burn them—and call in a consultant to compose a new set!

Joseph Hall, longtime chief executive at Kroger Company, has suggested that his people should periodically change their orientation to their jobs. In order to forestall loss of direction and overreliance on the past, "I would tell them that they should, a couple of times a year, pretend they are a new man on the job. All right, one of the first things you do is just go through your desk and look at the files and clean them out; and everything that crosses your desk, you question: Why? Never take things for granted. I don't believe in tradition unless tradition is sound."[6]

Loss of Spontaneity

Perhaps even more dangerous for the organization is when spontaneity disappears. Once established and accepted, policies and controls tend to limit flexibility and initiative. In the beginning at least, management gains in coordination and predictability what they may lose in initiative. Nonetheless, since policies, procedures, and controls are based on the past, it is difficult to keep them up to date: policies may no longer apply to new conditions, controls may measure irrelevant factors, and those rational plans which were developed to promote organizational effectiveness begin to interfere with the accomplishment of objectives. If managers blindly follow these rules, spontaneity is lost. Under long-standing restrictive control systems, some managers ignore whether mistakes are avoided or not. If they feel that punishment awaits any unsuccessful departure from procedure, managers may do what "the book" requires—and that's all. No spontaneity will be demonstrated and apathy will be prevalent. Bureautic behavior and bureau-pathology are descriptive terms which have been coined to illustrate such problems that endanger all organizations.[8]

Managers must structure organizations, allocate people to jobs, fill communication links, and determine operational objectives. Standing plans, policies, procedures, and performance standards are means to reaching these goals. Such codification of the past is necessary to maintain a sense of organizational direction and momentum. Yet, it can get out of touch with the real needs of the business; people may conform to rules without thought as to their contribution to goals so that the organization loses direction.

REFERENCES

1. Henri Bergson wrote: "We never experience our psychic states as simultaneous, existing alongside one another. Nor do we even find such states set off sharply from one another as discrete elements in the succession; they blend into one another in such a way that the present contains both memories of the past and anticipations of the future."
Harry Levinson indicates that "man can bind time. Not only can

he act now in anticipation of the future, but also the unconscious memories of unresolved problems which he carries continue to be alive within him as if they were yesterday's events. . . . People act in the present, using symbols, as if still living in the past with the real thing." (*Executive Stress,* p. 50)

2. On Ford, see Allan Nevins, with collaboration of Frank E. Hill, *Ford* 3 vols. (Charles Scribner's Sons, New York, 1954–1963), esp. Vol. 3, *Decline and Rebirth.* On General Motors, see Alfred P. Sloan, Jr., *My Years with General Motors,* edited by John McDonald with Catharine Stevens (Doubleday & Co., New York, 1964).

3. The eminent Columbia history professor Henry F. Graff once suggested that "Graduate students ought to have some sense that anyone who has written in their field before is dead wrong, if not maliciously so." But they better not be ignorant of those writers! Comments to the Columbia University Graduate History Society, 1963.

4. The examples of military inflexibility are discussed (along with some similar business disasters) by Elting E. Morison, *Men, Machines and Modern Times* (M.I.T. Press, Cambridge, 1966; copyright © 1966 by The Massachusetts Institute of Technology.)

5. On the inversion of means and ends, the comments of John Knowles, former Director of Massachusetts General Hospital, are illustrative: "In the teaching hospital, it has become set that the patient exists for the teaching programs and not that the hospital exists for the patient." Quoted in Jan De Hartog, "What Money Cannot Buy," *Atlantic Monthly,* Vol. 218 (July 1966), p. 113. For other examples, see: Chris Argyris, *The Impact of Budgets on People* (Controllership Institute, New York, 1952); Peter Blau, *The Dynamics of Bureaucracy* (University of Chicago Press, Chicago, 1965); Frank Jasinski, "Use and Misuse of Efficiency Controls," *Harvard Business Review,* Vol. 34, No. 4 (July-August 1956); and David Granick, *The Red Executive* (Doubleday & Co., New York, 1960.)

6. Joseph Hall, quoted in editors of *Nation's Business, Lessons of Leadership,* p. 209.

7. Victor A. Thompson, *Modern Organization* (Alfred A. Knopf, New York, 1961).

12
Future Time

Almost from the very appearance of homo sapiens, his handcrafted tools, burial arrangements, and fantastic cave art suggested concern for the future.[1] Ancient man took time out of the present to prepare physically and psychologically. This ability to draw upon past experience in the present "to plan for the future has conferred upon man an advantage which in the struggle for existence against other species has proven superior to their often greater muscular strength, swiftness of action or ferocity." Contrary to animals which are fully subject to their predetermined or accidental relations to time, man is granted the possibility—limited though it is—of conditioning the order of his time himself. The possibility: to relate himself freely to both his past and his future, and by means of these relations to mediate actively his present in the interest of that future. This is time's great promise.[2]

What ancient men did is not dissimilar to what managers do today. All human organizations from families to secret clubs to public institutions express man's recognition of time's passing coupled with a simultaneous desire for permanence. Chapter 1 suggested that the corporation is a particularly clear example of man's efforts to overcome time, to ensure a future, and to exploit it.

Beautiful thoughts, but awfully tough to act upon. For all managers are faced with the nagging reality of their jobs—pressure to concentrate on the present, ignore the past, and let tomorrow take care of itself. Managers, however, who are so taken in are not exercising human leadership to the fullest potential. Behaviorally and psychologically, management should be oriented toward the future, carving out of the present in the service of tomorrow—like those cave painters of old. "The ideal executive," maintains one corporation chairman, "should have a feel for the future, not only from the standpoint of where his own business is going, but where his competitors are going, what other materials are doing, and what the country is likely to do."

PLANNING

In A. A. Milne's childhood classic, *Winnie the Pooh,* young Christopher Robin tells his group of stuffed animal friends that before they can search for the North Pole, they ought to get organized. Pooh asks what getting organized means and Christopher replies, "Organizing is what you do before you do something, so that when you do it, it's not all mixed up." Of course, what the juvenile sage is describing is not organizing, but planning—the traditional management function for handling the future.

Much has been written about management planning and a thorough discussion is not appropriate in this book.[3] Nonetheless, some observations on planning, time, and the future are necessary. We shall first consider the nature of planning in this chapter and then discuss why inadequate time is allocated to it in the next.

The definitions of planning are many: thinking about what you want and how you are going to accomplish it; determining in advance what is to be done; preparing for the future by making decisions now; and so on. Planning represents management's attempt to anticipate the future and win some security against the menace of temporal change of which man's time sense makes him so urgently aware.[4] Planning then is our effort to visualize "the future as history" by determining how we would want the future to appear if we could jump out in time and look backward.[5] "The purpose of planning," according to one executive, "is not to show how precisely we can predict the future, but rather to uncover the things we must do today in order to have a future."

A Marxist economist once visited the United States and in talking to the press commented how strange American business was. He thought our managers were always complaining about the dangers of a planned economy, yet American businesses do more planning than anyone else in the world. The economist suggested that more time, effort, and money is expended on economic planning in the United States than in the Soviet Union. Of course, the planning may be more decentralized (and too short-ranged in perspective), but we do live in a planned economy and this is essential in a vast, industralized, interdependent system.

REFERENCE AND WISHFUL PROJECTIONS

So many books talk about planning as a decision process, but this is not most important. As my colleague, Russell Ackoff of the Wharton School points out, the key to planning is the determination of desires. The critical aspect of planning is knowing where you want to be and how you want the future to turn out.[6]

Central to planning—and lacking in many managers—is a sense of the ideal, a concept of where the organization should go and what it should become. Many managers have woefully underdeveloped imaginations about the future—as might be expected given the pressures of short-term concerns. In order to deal with the future through planning, two tentative plans must be drawn up even though neither is likely to take effect. The first has been termed a "reference projection." This is a prediction of what is likely to occur if management does not intervene, if nothing new is introduced.

The second projection is even more important, and more difficult. This is the "wishful projection," a definition of the ideal concept of where management wants to be and when it wants to be there. These aspirations have little likelihood of complete accomplishment and will probably change with time. Nonetheless, some sort of wishful and reference projections are essential if planning is to be more than ad-hoc responses to present conditions.

The wishful projection may have quantitative components regarding sales, market shares, profits, and so on, but the more important aspect is a qualitative scenario of the kind of organization management wants to create and the activity in which it wants to be involved. In short, what kind of business would the company most like to be?

The difference between these two unlikely projections—the reference and the wishful—define the gap to be filled. In practice, these first two plans tend to be skipped in the interest of next year's budget or, at most, the company's five-year capital needs.

Difficult as it may be for pragmatic, achievement-oriented managers to accept, some dreaming is crucial to future planning. Of course, such dreams should be more than a list of unrelated aspirations; the ideals should be integrated into a cohesive system.

Such wishful projecting has a threefold purpose: First, to provide a conception of what the organization should be like as a whole and ideally. Without this, even small correctional measures cannot be evaluated effectively. Even worse, lack of imagination, rather than excess of it, leads to mistakes and missed opportunities. Without a self-determined sense of the ideal, management tends to play follow-the-leader after the latest fad, whether it be operations research, theory Y, management by objectives—all worthy ideas but useless to the organization without a sense of its desired future.

Second, wishful projection may provide a guide for planning systematically for transition of the existing organization into what is desired. Without this, management will only proceed in piecemeal manner responding to current pressures rather than long-run needs. It is part of management's predicament, but a promise of the future, that we can often do least to change situations that are close to us and about which we know the most, while we may have greater influence over those future situations of which we know relatively little.[7]

Third, and probably most important, wishful planning may excite and fire imaginations of participating managers. "Planning," Ackoff says, "should excite managers to the point where they strain sufficiently at the leash of tradition and custom to break with them. I know of no better way to bring them to such a point than to involve them in the idealized design process."[8]

RETROSPECTIVE PLANNING

Some plans are less oriented toward the future than the past. This is retrospective planning. Of course, management engages in this activity in order to improve the future, for this planning is directed toward removing deficiencies produced by past decisions and reducing repetition of these mistakes. This is really systematic examination of the past. It is often difficult to do this because most people tend to forget the original reasons for their decisions. If possible, it helps to reconstruct the histories of how managers arrived at their present positions; then they know what they would be giving up if they change. In addition, it is important to understand the negative side of one's past choice, the costs and drawbacks that were associated with it. We should not forget that in any decision something has to be sacrificed or compromised in committing oneself to it. For example, in its formulation of yearly objectives, one utility company includes specific analysis of a sample of past projects to uncover weaknesses that they might not even have been aware of. At a more theoretical level, the necessity for retrospective planning underlies the need for an integrative management system to have a memory subsystem. This memory should receive a performance prediction for every decision which is retained for later comparison with actual performance.

PROSPECTIVE PLANNING

Prospective planning is more directly concerned with the future. However, management may have varying knowledge about the future which requires different kinds of plans:

• Management can know what the future will be with certainty—in which case, planning is mere *commitment* of resources. Most short-run production planning or response to governmental legislation approximates certainty and commitment.

• Management can be uncertain about the future—that is, aware of several possibilities and perhaps with some estimate of the likelihood of each. Here *contingency* planning is called for; planning that develops alternatives for each reasonable eventuality. This entails a certain amount of future prediction, but not in the sense that the future already exists through a glass darkly. Rather, managers estimate alternative futures, the likely results of different choices, so that they can understand the costs and consequences of different desires. This condition would exist when management lacks knowledge of whether and what legislation will be passed regarding quality standards, but does know what is being considered. Sometimes, we tend to be a little sloppy with contingency planning by concentrating only on likely possibilities. As Kahn and Wiener put it:

> Options should be provided not only for the important long-run choices which are most likely to occur but also for less likely choices insofar as they would present significant dangers or opportunities, and if preserving the options would not require

> giving up too much in current terms . . . we are not only concerned with anticipating future events and attempting to make the desirable more likely and the undesirable less likely. We are also trying to put policy makers in a position to deal with whatever future actually arises, to be able to alleviate the bad and exploit the good.
>
> Therefore, contingency planning hopefully will provide management with information: how situations might arise and what alternatives exist at each step for preventing, diverting, or facilitating.[9]

• Unfortunately, management may be ignorant about the future. But this does not mean that it is helpless. *Responsive* planning attempts to develop policies that take into account the conditions that may exist at the time action is required and that permit use of all relevant information available. For example, a policy that states that only college graduates will be hired for professional or managerial positions is wrong in *any* company. Aside from the fact that most managerial jobs just do not require such education, the policy is undesirable because it assumes an unlikely environmental stability. Economic recession or changes in educational philosophy and practice might well mean that the most promising junior managers would be college dropouts impatient to move from theory into practice. Some retail merchandisers already think this is true. Research suggests that such people have higher job morale and perhaps even performance.[10] Given the sorry state of urban public school systems, something similar might even be true for high school dropouts.

• Finally, there is *proactive* planning, where management tries to reduce its ignorance about the future and its environment by active intervention. Lobbying for certain legislation, changing public opinion, and supporting educational experimentation would be examples.

TIME SPAN OF PLANS

Another way of thinking about the relation of plans to the future is in terms of time span: long, intermediate, and short-range. When categorizing these time spans, most managers have been heavily influenced by the financial orientation that over five years is long-range, one to five is intermediate, and less than one is short. This is not a helpful scheme. Whether a plan is long- or short-range depends upon its impact on the business and whether or not decisions can be modified. A more appropriate approach considers the factors that management can manipulate. First, in long-range planning management can determine the structure and strategy of the organization. In other words, long-range planning may involve a fundamental redirection of the concern's ends and means. Second, in intermediate-range planning management must accept the general structure and strategy of the organization, but can manipulate the quantity of inputs and outputs, people, material, and capital. Third, in short-range planning, structure, strategy, and quanties are fixed, but management can manipulate the application of resources and technology in scheduling specific activities.

The time spans by each of these definitions vary widely. In the automobile industry, long-range plans would cover ten years or more, intermediate-range plans varying quantity and quality might be somewhere between one and two years, and short-range plans might include two to four weeks. Contrast this with a small garment manufacturer doing business in a rented loft. Its long-range plans can extend probably no more than six months to a year because it is so heavily dependent on the styles and ideas generated by designers and customers elsewhere. Intermediate plans would run two to three months, and short-term plans would be on a half-day basis. Therefore, the difference between long- and short-range plans depends on the organization's technology and product. But, all three time ranges are necessary. Even though its horizon is closer in time, the Mod Clothing Company needs its six months' long-range planning as much as General Motors needs its ten-year perspective.

The major problem in planning over any time span, of course, is inflexibility. The ill-fated Edsel provides the classic story.[11] When the idea for the Edsel originated in the early 1950s, Ford wanted to fill a gap in the middle of its product line in order to compete more favorably with General Motors in this attractive profit margin area. When the first designs were formulated in 1954, the automobile industry was entering the era of longer, lower, heavier, fancier, and gaudier models. Extensive market surveys were taken and the conclusion, rightly or wrongly, was that the public would purchase a new automobile created on these lines. Consequently, the original Edsel was one of the biggest and flashiest cars in history. From the basic design to the first model required approximately three years; the first cars came to the market in the fall of 1957. They apparently did fairly well initially, but then the market collapsed. The Edsel lingered near death for two more years before it expired.

The explanations for this debacle are many and speculative. The simplest is that the public did not like the design, especially the grille (suggesting that the sexual imagery on which the design was supposedly based was met easier by other media). But, perhaps a more pertinent reason was the unfortunate coincidence of its birth with other developments. In October, the Soviet Union orbited the first artificial satellite. This shocked the United States and produced a serious reexamination of American goals and strategies in the nuclear age. For a time at least, we turned toward more austere and conservative cars—a trend strengthened by economic recession in 1958 which gave a boon to American Motors and the compact automobile. Edsel's concept and design may have been correct for its assumptions, but the world changed and the initial idea was no longer relevant. This is the danger in all planning.

In summary, the elements of planning include at least:

- Reference projection
- Wishful scenario
- Retrospective analysis
- Prospective planning
  —commitment decisions
  —contingency planning
  —responsive policies
  —proactive plans

What kind of planning is good or bad is not the issue as much as balance. The worst situation would be exclusive concentration on only one type of planning. This would most likely be just retrospective examination of past mistakes, a witch-hunt for villains. Much less likely, but equally bad, would be drifting with wishful scenarios without any prospective plans for implementation. Proactive planning of environment is obviously the most difficult, especially in a turbulent era. A balanced effort to include all types is desirable. Long ago, Machiavelli argued that half of men's actions are ruled by chance, and that the other half are governed by men themselves. The actual ratio is unknown, but rational consideration of the future through planning reflects management's effort to modify the relationship.

It's all so clear. Managers must plan; they must separate themselves from the burdens of the past and the exigencies of the present in order to define desired conditions in the future. Nonetheless, apparently many do not do so. Executives tend to agree that planning is their most neglected duty. Why? Mainly because the present dominates their allocation of time. We shall examine this in the next chapter.

REFERENCES

1. On the significance of early cave art, see S. G. F. Brandon, "Time and the Destiny of Man," in Fraser, *The Voices of Time,* p. 140.
2. Friedrich Nietzsche described man as "the only animal who can make promises." The best survey of human nature and its relation to time is Reinhold Niebuhr, *The Nature and Destiny of Man* (Charles Scribner's Sons, New York, 1949).
3. The most comprehensive treatment of planning is probably George A. Steiner, *Top Management Planning* (The Macmillan Co., New York, 1969). A very fine small book is Russell L. Ackoff, *A Concept of Corporate Planning* (Wiley-Interscience, New York, 1970). Much of my discussion of planning, including the general categories of plans, follows Ackoff. Some of Kahn and Wiene , *The Year 2000* is also very helpful for understanding planning.
4. "In a true sense, all the planning that must necessarily precede each human enterprise represents an attempt to anticipate future contingencies and win some form of security against the menace of temporal change of which man's time sense makes him so urgently aware. But the advantages which this time sense thus confers in material benefits are offset by what may be described as a profound sense of spiritual insecurity which also stems from consciousness of time." (Brandon, *op cit.,* p. 144)
5. See Robert Heilbroner, *The Future as History* (Harper & Row, New York, 1960).
6. Here is how Rollo May describes man's potential: "The human wish . . . is not merely a push from the past, not merely a call from primitive needs demanding satisfaction. It also has in it some selectivity. It is a forming of the future, a molding by a symbolic process which includes both memory and fantasy, of what we hope the future will be. The wish is the beginning of orienting ourselves to the future, an admission that we *want* the future to be such and such; it is a capacity to reach down deep into ourselves and preoccupy ourselves with a longing to change the future . . . it is not wishing which causes illness, but *lack of wishing.*" (*Love and Will,* W. W. Norton & Co., New York, 1969, pp. 211–215. Copyright © 1969 by W. W. Norton & Company, Inc. Reprinted by permission of the publisher.) This may be true of American society and business as well. The few executives who are conspicuous for their capacity to express dreams for the future and to define great missions (such as Irwin Miller of Cummings Engine and Joseph

Wilson of Xerox) unfortunately are rare. The limited imaginations of many managers is disappointing. Harry Levinson cites a survey of managers that reports that they tend to define the future in terms of their present experience: they do not predict much change over a period of time. They seem to say, "What is, will be." (*The Exceptional Executive,* p. 169)

7. A point made by Kahn and Wiener, *op. cit.,* p. 2.

8. Ackoff, *op. cit.,* p. 63. (Copyright 1970 by John Wiley & Sons, Inc.)

9. Kahn and Wiener, *op. cit.,* p. 3. (Copyright © 1967 by The Hudson Institute, Inc.)

10. For an extensive discussion of education, morale, and performance, see Ivar Berg, *Education and Jobs: The Great Training Robbery* (Frederick A. Praeger, New York, 1970).

11. See John Brooks, *The Fate of the Edsel and Other Business Adventures* (Harper & Row, New York, 1963).

13
The Dominance of the Present

Men allow the present to dominate the future and curtail planning for several reasons: they fail to think about the future, for it is just not real to them; perhaps they think about it, but consider it less than worthwhile to do anything; or punishment for immediate mistakes is just too close to worry about long-run failure.

NOT THINKING ABOUT THE FUTURE

The virtues of an achievement orientation can also be vices. As discussed in Chapter 2 managers with high achievement need desire feedback indicating they have performed effectively. But they tend to be impatient. They want results quickly, for that is the purpose of their effort and discipline. Unfortunately, overconcern with rapid progress and feedback hinders progress. Too much attention is directed to the present, to the short-run payoff, without adequate concern for the future. In the effort to act now, managers fail to collect necessary information, to plan, and to think.

Many observers suggest that this predilection toward action over thought is characteristic of American culture in general. Patience is an easy virtue where the prevailing cultural view of time is circular, with its infinite length and repetition. Patience, however, is difficult with a linear perspective. We are concerned about the future, but our future must be relatively close. Ours is not the distant future of the East, which is far beyond present influence; ours is just up the road a piece. Witness the situation in Indochina: for years we deluded ourselves that victory was around the bend, or that we saw the light at the end of the tunnel, or that it was the beginning of the end. Yet, the enemy's time scale for accepting warfare and delaying victory was beyond ours. "While we look to the future," Edward Hall writes, "our view of it is limited. The future to us is the foreseeable future, not the future of the South Asian that may involve centuries."[1] And that was written years before our military involvement in Vietnam.

Not thinking about the future may even stem from being successful in the present and confident about the future. The most pervasive theme in James MacGregor Burns's biography of Franklin Roosevelt is short-run ingenuity and tactical brilliance.[2] Apparently the very competence of the President to manage the turbulence about him led to enormous confidence that he could similarly manage the Soviet Union and postwar conditions. No real plan or even future philosophy was formulated to implement this confidence—and unfortunately F.D.R. died. Such conditions of competence in the present and perhaps overconfidence about being able to handle the future may characterize many successful managers. Therefore, they never find the time to consider the future. As former R.C.A. chairman David Sarnoff suggests, "You know our brain needs exercise as well as our muscle. A great many people don't go beyond today. If things are going along all right today, they are not thinking about five, or ten, or twenty years from now."[3]

Not thinking about the future finds its most common form in inability to defer gratification. To lower forms of life, *now* is the only reality; needs must be satisfied immediately. Higher up, the intervals of gratification delay increase: we are told that a rat can sustain a delay of some four minutes; a cat, seven hours; a chimpanzee, forty-eight hours. In man the horizon may reach beyond his own existence. Of course, people vary. Children have short time-delay spans; it is difficult to postpone satisfaction because the future lacks much meaning to someone without much of a past. Immature persons act on the basis of immediate need without much anticipation of later, secondary consequences. So high insurance rates for youthful drivers reflect not their ability (which is superior to older males), but their tendency to consider only the present when tearing along the highway heedless of possibilities around the bend.

What is attributed to youngsters is characteristic of most adults in some cultures, and perhaps everyone everywhere under certain conditions. Anthropologists have observed that for some Indian tribes, only the present has reality.[4] Similarly, in countries whose ancient glories are celebrated as if they happened yesterday, businessmen have been known to invest large amounts in factories without making the slightest plan how to use them. In this case, the future seems unreal.

LACK OF FAITH IN THE FUTURE

One of the hoary traditions of white America is the myth of the happy-go-lucky darkie eating watermelon, dancing to his jazz, without care or thought for the future. The myth's persistence ("They'd be happy if it weren't for outside agitators") suggests that whites find time burdensome—and that they long to be freed from its concern. But, or course they cannot and neither can their black brothers. For inadequate thought about the future has not hurt America's blacks so much as thinking about it and not seeing much there—not much at any rate that they can do anything about *now.* A bird in the hand, cross that bridge later, and the poet's "take the cash and let the credit go, nor heed the rumble of a distant drum" are all attractive to all men sometimes, but hold special appeal to those who doubt what is in the bush, expect not to ever reach the bridge, or know from experience that promises tend to be broken—in short, to those with no parents, relatives, or friends for whom any other perspective has paid off.[5]

Although such conditions frequently apply to the poor, there is no intrinsic connection between present time orientation and lower class. Whenever people of any social class have been uncertain or downright pessimistic about the future, "I want mine right now" has been a common response. The legends of promiscuity during World War II at some colleges dwarf the tales about today's students. In stressful times all classes tend to throw off conventional restraints and appear less able to defer gratification; future time orientation appears to collapse toward the present.

There is a story about an American agricultural expert visiting in the United Arab Republic.[6] The American asked a local farmer about his expected yield. The Arab was highly insulted since he thought the American considered him crazy. To the traditional Arab, only God knows the future and it is presumptuous even to talk about it. Perhaps the Egyptian farmer had faith, but planning requires greater faith because the expenditure of time now is an investment in the future—and sometimes a risky investment. It assumes that tomorrow will come and that action today will make a difference. It certainly presumes giving a damn.

So people may not plan because they fail to think about the future or because they can conceive of nothing good there. Individually most managers are achievement-oriented, time-conscious, and mildly optimistic—at least to the extent of believing that action helps. Within organizations, however, ignoring the future is not rare. Much talk is heard about the future in business and politics, but until recently the talk was preceded by little real thought—witness the reluctance to think about the implications of despoilment of the atmosphere, rivers, lakes, and oceans. Inability to delay gratification has certainly characterized us here.

Not thinking, lack of faith, and impatience are important reasons for the dominance of present over future. The most important reason, however, is the immediacy of failure if present problems are not managed effectively.

PROXIMITY OF REWARD AND PUNISHMENT

The famous economist John Maynard Keynes once acidly responded to the conservative's insistence that in the long run economic depression would cure itself through the free market. His oft-quoted reply: "In the long run, we're all dead." For Keynes, and for President Roosevelt eventually, saving private enterprise and democratic society were more important than preserving a certain model of capitalism. Perhaps they were right, perhaps not; the problem is that the manager cannot wait. Without action, failure looms.

Politicians and government officials tend to concentrate on the short run because of threats in the near future: defeat of a bill, reduction of the yearly appropriation, rejection by the voters.[7] Long-range demographic, agricultural, or environmental problems will not hurt them *now* (although they may not be able to say this much longer). In this short-range view, small matters bulk large.

Political history suggests that a manager needs one thing above all others to accomplish great and lasting things—time. Conquest and destruction can be completed quickly, but building takes longer. Consider the United States government's apparent repetition of so many of France's mistakes and even our own earlier errors in Southeast Asia. The sense of *déjà vu* is overwhelming when looking at articles of twenty, even five, years ago. A Rand Corporation researcher comments:

> Through the years, the same strategies have been used with different names, but the same old mistakes have been made. Why have these mistakes been repeated? For one thing, there is a lack of institutional memory in the American organization in Vietnam. Any lessons learned here have not been transmitted to people who followed the people who learned the lessons. This is due primarily to the short tours of duty, the rapid turnover of people. The one-year tour—especially for people in command jobs or running programs—has been disastrous. A person comes in, works hard for a year and then leaves, giving the job to someone else with no experience.[8]

Life and death are not as close to the business manager, of course, but dominance of present over future can be as great. Short-term performance standards encourage many managers to concentrate on the *now*. Frequently they feel that they are rewarded or punished for *this* year based upon short-run, objective measures of profits, costs, and growth. In the long run they will be dead—or transferred. A competitive business system encourages a short time perspective of course; only a monopoly can ignore the present. Yet, even given competition, management tends to measure performance over too limited a time span. Concentration on the present is sometimes rewarded, while concern for the future is punished.

Corporate rotation and transfer policies also contribute to the dominance of short-range time orientations. Supermobile managers average a position change every eighteen months; even average managers move every two to three years (and presidents may even remain in office for shorter periods).[9] Transfers and shorter tenures do not reflect job mastery and threat of stagnation so much as they express impatience: not only impatience concerning performance feedback, but also indication that one is progressing up the hierarchy. Mobile managers tend to measure their success by how frequently they are promoted/transferred, not by how well they perform. They never are afforded the luxury of failure or time to learn, so the first setback is catastrophic. This lends credence to Laurence Peter's *Peter's Principle* that every managerial slot in hierarchical organizations is eventually filled with an incompetent because managers are promoted to the level of their incompetence. This is an exaggeration, of course, because men can and do grow as they move upward, but the point is clear: our national culture, organizational climate, and personal attitudes emphasize the present.

REFERENCES

1. Edward T. Hall, *The Silent Language,* p. 20.
2. James MacGregor Burns, *Roosevelt: The Soldier of Freedom,* (Harcourt Brace Jovanovich, New York, 1970).
3. Quoted in *Nation's Business,* June 1966.
4. For an interesting book on Indians and their time perspectives among other things, see Peter Farb, *Man's Rise to Civilization as Shown by the Indians of North America* (E. P. Dutton & Co., New York, 1968).
5. In his marvelous book describing the society around a single ghetto street corner in Washington, D.C., Eliot Liebow describes the future facing a young man named Richard: "It is a future in which everything is uncertain except the ultimate destruction of his hopes and eventual realization of his fears. The most he can reasonably look forward to is that these things do not come too soon. Thus, when Richard squanders a week's pay in two days it is not because, like an animal or a child, he is 'present-time' oriented, unaware or unconcerned with his future. He does so precisely because he is aware of the future and the hopelessness of it all. . . . [Richard speaks] 'Maybe I look like a boy and maybe I act like a boy sometimes but I got a man's mind. You and me don't want the same things out of life. Maybe some of the same, but you don't care how long you have to wait for yours and I-want-mine-right-now.' " (*Tally's Corner,* Little, Brown & Co., Boston, 1967, pp. 64–68)
6. The Egyptian story is told by Hall, *op. cit.*
7. For an insightful consideration of the pressures on government officials, see Frankel, *High on Foggy Bottom.*
8. The observations on Vietnam are by Gerald Hickey, an American who has lived in Saigon for the past 15 years. Quoted in *Newsweek,* May 3, 1971.
9. On trends in mobility patterns, see Eugene Jennings, *The Mobile Manager* (University of Michigan Press, Ann Arbor, 1967). Another reason that managers (and even their wives) may like fairly frequent moves is that they create a period when a "short-timer's" attitude is permissible and quasi-legitimate. It is a vacation of a sort, a time of luxury, because one need not be concerned about even the near-future of one's present unit. Off-the-job, a similar attitude can prevail; one need not worry about the community or feel much compulsion to get involved because it is not your community and you will not be there very long.

14 Reconciling Past and Future

What exists at any present time is tension between past and future, between paradoxical need for both stability and change. This tension is especially dramatic in the historical struggle between authority and freedom, control and autonomy. Although stressful and dangerous, we need this struggle. Numerous political observers have suggested that we should maintain balance in the contending forces. Some like Hobbes, Bodin, and Lenin (and possibly Brezhnev) were inclined to have the balance skewed toward authority and control. Others like Locke, Jefferson, and Kropotkin (and perhaps Dubcek) emphasized freedom and autonomy. Long ago John Stuart Mill indicated the requirement for creative tension when he wrote: "A party of order or stability, and a party of progress or reform are both necessary elements of a healthy state of political life."

So it is with society's organizations. Business has emphasized control and authority for management may owe more to Ghenghis Khan than to Thomas Jefferson.[1] Given economic realities, the emphasis may not be misplaced. Unrest exists however. Many people call for reexamination of existing authority systems in all kinds of organizations to see if control has been converted from a means to an end in itself.

STABILITY AND PREDICTABILITY VERSUS CHANGE AND SPONTANEITY

The tension is expressed in today's confrontation between stability and change, predictability and spontaneity. Youth with their "now" sound in music, their "pop" art form, and "sock-it-to-me-baby!" life style cherish sincere spontaneity in human relations. In politics some favor change without apparent awareness of what is more desirable. A small minority are vocal in telling us that uncontrolled change—chaos—is preferable to the status quo. "There is a certain relief in change, even though it be from bad to worse." So wrote Washington Irving two hundred years ago.

In fact, every organization requires both change and stability. Change is exciting, so much literature is devoted to the manager's responsibility to initiate change. But stability is just as important. Most of the time, most administrators are trying to establish and maintain equilibrium in workflow and human relations—not because they fear change, but because individuals and systems can only absorb limited innovation. For effective relations and efficient performance, relative stability is essential. Without such stability, the advantages of division of labor will not be realized, cooperative human relationships are hindered, and organizational control is handicapped.

In any organization jobs are interdependent; people must be able to predict the behavior of others in interacting positions. Whether on the engine block casting operation at Chrysler, or the policy underwriting unit at Liberty Mutual, or even with the Students for Democratic Society "liberating" a university building, coordinated effort will collapse into confused individual actions unless behavior is highly predictable.

Stability and predictability are also necessary because of human limitations. We are *ceteris paribus* creatures; we can consider only a relatively few factors at a time. Other factors must be assumed constant. This means that we must focus our attention. Research on the selective nature of perception indicates that we do so by concentrating on matters that are most central to our needs and ignoring those that are peripheral.[2]

Most important, novelty and creativity are unrecognizable unless they emerge from order. Laboratory research on group dynamics supports this. In solving problems and adapting to external changes, it appears that the whole group should not change behavior, only some members. Most effective adaptation seems to occur when some members repeat past practice while others change behavior completely.[3] Thus, a mixture of repetition and change may be the best way to make the transition from past to future. This implies that in order to behave adaptively, management needs stable regulations and procedures that it can employ in carrying out adaptive procedures. That these rules are overdone sometimes is true, but the three-hundred-year-old advice of one Lucius Cary still has relevance: "When it is not necessary to change, it is necessary not to change."

PROMISE OF THE PAST

Contrary to the examples in Chapter 11 and to loud cries among some of the younger generation, the past is not all bad. A continuing problem for all societies has been that revolutionaries and radical reformers tend to measure progress by the number of antiquated institutions that are cast off. Those who favor progress tend to simple-mindedly believe that evils are the property of the past and that only by moving away from the past as rapidly as possible can a better future be assured. But, they are wrong. As an anthropologist points out:

> It is often assumed that decay is due to the dead hand of conservatism, and it is of course the fact that religious and political theories which involve a belief in the infallibility of the ancient mind often lead to decay. . . . It is less often realized, on the other hand, that decay of culture can be brought about even more rapidly by breaking away from the past.[4]

A sense of respect for and continuity with the past are essential for a viable institution. Historical continuity with the past, according to Justice Oliver Wendell Holmes, "is not a duty, it is only a necessity." As a nation we are experiencing difficulties partly because of inadequate understanding about our past and our inability to integrate legitimate past values into the present. The United States tends to be historically blind. This is unfortunate, for as John F. Kennedy once observed: "History . . . is the memory of a nation. It is the means by which a nation establishes its sense of identity and purpose." Our youth, our rejection of Indian history, our creation of the new, our growth and expansion have placed emphasis on the here, not the then. Contrast this with Mexico's acceptance and building upon Aztec traditions. Time in the United States seems to be a commodity, but only present time. The past is useless (except in western movies that provide simplistic escape from today's complexities), the future of interest only as a potentially better present (or to demonstrate through science fiction that it may be even worse).

The problem in the United States is that we refuse to confront the past except in deluding myths. For example, consider our attitudes about firearms. We have been unable to deal with the reality of violence and guns in our history. Public opinion polls suggest that we still believe that violence is the best way to overcome violence (including shooting looters) as well as believing that possession of guns is the "great equalizer" of the old west—the trusty sidearm of the Wyatt Earp who confronted the villain man-to-man. But, we delude ourselves even in our history. Marshal Earp was a passionate believer in disarming America. His style of law and order was to confiscate the arms of citizens and visitors to his community. He would probably be aghast at what we allow in our complex society today. We ignore or distort the past to suit present desires.[5]

In large corporations, orientation sessions, booklets, and movies provide some introduction to the company's past, but these tend to be done badly. Saccharin and Pollyannaish, they imply that virtue was rewarded. They leave out the blood, sweat and tears—and probably the conniving—that went into that past. Every new manager ought to have some knowledge of the times that his predecessors should have gone to jail. Company histories, even public relations releases, need to be rewritten as school texts are being rewritten to tell it as it really was.[6]

Managers should be time-binders.[7] The hierarchy of differentiated functions, power, and status characteristic of management is not unique to mankind. Apes, quadrupeds, and multilegged creatures of various sorts demonstrate such structures. Many of the principles of management apply equally to baboons—only these cannot read or talk about them. To reflect and promote uniquely human attributes, therefore, organizational management must demonstrate a time-binding role, the unifying of past and future through the present. Such melding of past, present, and future helps give order and predictability, promotes individual commitment, and lessens the need for those bureaucratic rules under such criticism today.

A time-honored way of binding past and present has been through ceremony and ritual. Academic institutions abound with such events—colorfully garbed actors participating in convocations, commencements, baccalaureates, "cane sprees," "heydays," and so on. Similarly, the military promotes ceremony in reviews and assemblies, commissionings, decorations, and changes of command. So also the church. Unfortunately, many other institutions do not. Where business has, it is now thought to be corny or childish. Yesterday's I.B.M. and Metropolitan Life beanies and songfests are unacceptable to modern tastes. Even the twenty-five-year dinner and watch are suspect. Their passing is not mourned, but the need remains. Even in academia, students and faculty are turning away from the old ritual; seniors often must be threatened to show up for graduation! Yet, the need for time-binding ritual is not dead in schools, churches, or in business. It may be stronger than ever,

Consider the widely publicized anti-Vietnam War Moratorium demonstrations in which thousands of young people (and some older ones too) participated. To this observer they seemed less "anti" demonstrations than communal rituals, rather like old-fashioned religious processions and services. To be sure, these were not historical ceremonies; emphasis was on the present and the need for change, but they suggest the necessity for such ceremony—especially in an era when scouting, school, and even church no longer fulfill man's need to feel part of an historical community coming from somewhere and going someplace.

Management's problem is how to appeal to his need. With today's sophistication and pseudosophistication, any condescension is resented. Open appeal to emotion is also rejected—especially when the goal fails to justify lofty rhetoric. A number of young former M.B.A. students (now in business) have voiced contempt for even moderate emotional appeals from their superiors to sell more Admiral Zoom cereal or flowered toilet paper. Yet, time-binding ritual is necessary; creative manager-leaders must search for ways.

Company executive programs provide one limited possibility. The content may be less important than the fact of the program and the composition of the group. Bringing people of various ages together will almost force history into the discussion. Attaching prestige to being selected will buttress the modest amount of ritual inherent in the process. But, this is not enough; more is to be hoped for.

A sense of balance is obviously needed. The past must be respected, but not worshipped; it must be known, but not copied. John Kennedy was fond of paraphrasing George Santayana's observation that those who do not know history are doomed to relive it. Of course, history never repeats itself. Repetition is less the concern than sensitivity to trends, the big picture, or the view from the mountain. A few years ago a wealthy old graduate donated $5,000,000 to his alma mater. At a dinner honoring him for his gift, he was asked how he did it, starting more or less from scratch financially, when the days of the fortune-makers are supposed to be over. The generous investment banker responded: "In this I can say that History has helped a great deal, the long view rather than the short view. It is the waves which count rather than the ripples; the tides in man's affairs rather than the waves."[8] (After voicing his stirring words on the uses of history, he went on to add that he had never had a course in economics or been to a business school!) The point is that Americans tend to overlook the large patterns of history. Our horizons are so close that we ignore that which is just a little farther out—too far for us, but not too far for others with a different sense of time.

The message here is the relevance of the past to the present. In his prizewinning *Management and Machievelli,* Anthony Jay uses historical analogies in discussing management. He quotes the well-known British historian, Hugh Trevor-Roper, who writes: "Historical parallels can never be drawn without risk, but general lessons can be extracted even from societies distant in time or in place." And Jay adds, "There are far more lessons, for those who care to read them, in the long annals of history than the few published case studies."[9]

PROMISE OF THE FUTURE

The Queen told Alice in Lewis Carroll's *Through the Looking Glass* that it was a great advantage to have a memory that worked both ways. Alice replied that she was sure her memory did not work in that fashion, that in fact she could not remember things before they happened. "It's a poor sort of memory that only works backwards," the Queen retorted.[10] So perhaps it is a queer sort of history that only has relevance to the past. Our concern with the past should be in the service of the future. Although the past can be rewritten (and should—George Orwell's *1984* notwithstanding; witness the need to rewrite American history to include blacks, Indians, and Spanish-speaking Americans in their rightful and actual roles), we rewrite only to bring about a better and more truthful future. For the "one-wayness" of time means that man's essential freedom is creation of the future, not the past.

Consider the following example in which Russell Ackoff describes how one company imaginatively drew on the past and then current trends to shape its place in the future:

> In the process of developing a long-range plan for an appliance manufacturer, attention was directed at the question, Why were appliances developed in their current form? That is, why had particular functions (e.g., cooking, food preservation, and dishwashing) been isolated and mechanized? Reconstruction of the past showed that appliances had emerged during a period in which consumers had little capital to invest in them, and hence designers had concentrated on mechanizing as few functions in combination as possible in order to produce low-cost units.
>
> However, examination of buying behavior in the current market revealed that two or more appliances were usually purchased simultaneously. Such purchases are usually associated with remodeling or a move. This suggested combining functions in general-purpose appliances. Analysis revealed considerable cost savings in doing so.
>
> Generalization of this idea led to the design of a kitchen as an integrated unit, as a mechanized system. Such a "product" was developed and is currently available. The bathroom was similarly examined, with similar results.
>
> Still further generalization led to consideration of the possibility of manufacturing a house, not prefabricating it, but manufacturing a complete unit. A study of future housing needs, distribution of income, and construction costs made the future market for such houses look quite attractive.
>
> The plan emerged with a transitional schedule from appliance manufacturing through room manufacturing, to house manufacturing. It involved entry into the house-trailor business because the mobile home is already a manufactured unit that enjoys the largest share of the low-cost housing market.
>
> One further generalization was made. The community itself was examined as a potential product. Analysis revealed the growth of a market for new communities and the possibility of considerable sales and profit from them.
>
> Within ten years after completion of the plan, the company involved is profitably engaged in each of these extensions of the appliance business and is becoming a major force in reshaping the rooms, houses, and communities in which we are and will be living.[11]

True freedom is the opportunity to escape from the passing moment and the present pace, to challenge past experience or modify future action. Total freedom, however, is an impossibility for any manager or organization because of demands, interdependencies, and history. The appliance firm built on its past; it did not depart from it to produce automobiles or bake bread. But it did exercise some freedom in anticipation of the future. All managers should strive for at least small handfuls of true freedom. Without it, the future leads to obsolescence and inadequate development.

This is a special danger for young managers and specialists. Their immediate job demands interfere with such self-renewal. There are too many short-term pressures and too little time to do the amount of outside reading and studying that many young men should do for their professional growth. Does industry give sufficient time or incentive for young men to prepare for tomorrow? What can be done to provide time and a place in job descriptions so that the men who show the best prospects for advancement can have more opportunity to read, to take additional training, or to reflect on their experience? The demands placed on the best young men may in fact leave them less time for reflection and study than some of their less promising associates.[12] Since they are achievement- and action-oriented, they tend to value action over contemplation and to fill up passing time with the present. If the organizational climate encourages such behavior, the stage is set for later disillusionment.

Training programs, tuition refunds, company libraries are all familiar, but it is probably more important that self-development be integral to the job. Unfortunately, many employees and managers feel that study on the job must be sneaked, that somehow it is goofing off, or worse, that having the time to study indicates that they are little in demand and not highly thought of.[13] Some believe this because it fits their own action bias and their achievement-oriented sense of time.

All managers should think of themselves as educators.[14] They should build teaching into their daily activities. The approach to management then becomes less "What do I have to tell my subordinates?" and more "How can I help my subordinates to learn?" The latter is more future-oriented than the former because it should facilitate thought about now and then, where they are and where they want to go. If managers organize themselves for teaching while managing, they will grow themselves. Teaching will compel them to reflect upon what they are doing as well as upon the nature and problems of the business. Stimulated by the need to convey their thoughts, and to respond to subordinates' questions, they will think more. Time devoted to development is (and probably always will be) only a small part of the whole. In one study of executives, however, it was noticeable that those who spent the most time working alone also were those who during their meetings with subordinates asked the highest percentage of development questions.[15]

NEED FOR SACRIFICE

To express it in old-fashioned terms, sacrificing some present time and present satisfactions for the future is critical. Many of us in the United States seem to have lost any readiness to forego immediate rewards in the hope of promoting a more desirable future. And this is true of both managers and workers. Greater inclusion of workers in management might be a helpful vehicle for extending their perspectives at least up to management's. Numerous labor leaders have grown to understand the time dimensions of corporate problems as they have engaged in dialogue with management about costs, profits, and losses (unfortunately, many such leaders are then turned out of office for getting out of touch with their constituents). The American Scanlon plans and European workers councils offer possibilities for involving workers and stretching their perspectives although the experience so far has not been entirely encouraging.

Are managers beginning to bridge the gap between past and future more effectively? Perhaps. Over twenty-five years ago, a pioneer survey of top management practices by three Stanford University researchers observed:

> There is nothing about an organization more important than its future. Owners, management, employees, and society in general are, or should be, more concerned about where a company is going than where it has been. In any institution the responsibility for visualizing, initiating, and achieving future objectives rests with top management. The more specifically the future course of a company is conceived and defined, the more likely is its realization. One of the greatest needs observed during the course of this study is for more adequate planning and clarification of future objectives, both near-term and long-range.[16]

Recently these same corporations were revisited by the researchers who commented on expanded attention to planning and concern for the future. According to their report, chief executives in all firms now devote from 25 to 70 percent of their time to long-range planning (this is probably a generous estimate, however). Along with this top-management attention has occurred a decline in the actual planning performed by staff planning departments and an increase in such activity by line managers. This is a hopeful sign because it suggests a lengthening of perspective by those most under short-term pressures. "Currently," the researchers point out, "it is unmistakably evident that the executives who determine the destiny of their particular company are fully conscious of the inescapable obligation to plan for the future."[17]

Growing concern about ecological issues may reflect awareness of the escalating tension between present and future. With respect to our planet, we have managed our organizations much like the inch-worm in the oak tree; he unceasingly gorges himself raining his prodigious wastage on those unwary enough to wander below. If the bugs are few and the tree large, it is messy but not fatal. Unfortunately, we are many and our tree is not so big—as television transmissions from space have shown so dramatically. Nowhere else would management's abdication of its central human role of binding past, present, and future produce more tragic results.[18]

Man's gift of time consciousness is not an unmixed blessing. Fear and anxiety go with awareness of the future. As Kierkegaard wrote: "Time does not exist without unrest; it does not exist for dumb animals who are absolutely without anxiety." The material benefits that man's time sense facilitates are perhaps offset by a "profound sense of spiritual insecurity." As societies and organizations become more complex, time takes on ever greater importance. Concern for the future is more critical because future dangers are greater—for man than for animals, for urban man than for rural man, and for human society than for individuals. The stakes in the struggle between permanence and transitoriness become higher. The rewards of success and the penalties of failure become greater.

REFERENCES

1. The comparison to Genghis Khan was made by W. H. Ferry, *The Corporation and the Economy* (Center for the Study of Democratic Institutions, Santa Barbara, 1959), p. 2.

2. See James March and Harold Simon, *Organizations,* pp. 169–170; and Harold Simon, "The Architecture of Complexity," *Proceedings of The American Philosophic Society,* Vol. 106, No. 6 (December 1962), p. 476.
3. Research cited by Karl Weich, Professor of Psychology, University of Minnesota at an Eastern Academy of Management meeting, May 1971.
4. The passage from Raglan on the dangers of departing too rapidly from the past is cited in Lewis Mumford, *The Myth of the Machine: The Pentagon of Power* (Harcourt Brace Jovanovich, New York, 1970). This need to control change and conserve that which we value is a major theme in Mumford's work. He writes: "For the sake of continuity and cultural accumulation, previous cultures have usually over-valued custom and tradition and have even preserved their errors lest in extirpating them they forfeit their achievements. But the notion that the past, instead of being respected, must be liquidated is a peculiar mark of the megatechnic power system." (p. 282)
5. To quote once again from Mumford on the need to confront the past: "Western man, in his growing self-confidence, wanted to wipe the slate clean and begin all over. And therein lay a trap; for in order to overcome time, in order to begin anew, it was imperative for him not to run away from his past, but to confront it and literally to live down its traumatic events within himself. Until every generation did this consciously, examining its hoary tradition in the light of new experience, evaluating and selecting every part of its heritage, man could make no fresh start . . . So it remains an urgent task for our own day." (p. 14)
6. On corporate history, for example, there are two books on Thomas J. Watson and I.B.M.: The "official" biography, Thomas G. Belden and Marva R. Belden, *The Lengthening Shadow* (Little, Brown & Co., Boston, 1962) and an unofficial study, W. H. Rodgers, *Think* (Stein and Day, New York, 1969). The second book seems much more realistic and helpful in understanding the nature of I.B.M., its present and possibly future problems. Yet, apparently the company actively opposed Rodgers' work.
7. This need for managers to play a time-binding role in using symbols to integrate past and present is pointed out by Philip Selznick, *Leadership in Administration: A Sociological Interpretation* (Harper & Row, New York, 1957).
8. Shelby Cullom Davis donated $5,000,000 to Princeton University in 1965. See his "The Uses of History," *Princeton Alumni Weekly,* November, 24, 1964, p. 10.
9. Jay, *Management and Machiavelli,* p. 28.
10. The quotation from *Through the Looking Glass* is the theme for Morison's *Men, Machines, and Modern Times.* He cites it just after a discussion of antisubmarine warfare in World Wars I and II (p. 68).
11. Ackoff, *A Concept of Corporate Planning,* p. 57. (Copyright 1970 by John Wiley & Sons, Inc.)
12. The observation that the better young managers have even less time

for preparation and development for the future is made by Dill, *et al.*, *The New Managers,* p. 243.

13. On excessive workloads, Sune Carlson writes: "There is a definite need at the present time for industrial leaders with a wide knowledge of current social and cultural affairs, and it is impossible to acquire and maintain such a knowledge if one must spend all one's free time reading technical literature and business reports. There is also another aspect of the problem. It is only natural that the chief executive's way of working should become the standard for his subordinates. Instead of considering work on overtime as something exceptional, the people around the chief executive are led to regard it as the rule, and instead of widening their general cultural education, they specialize themselves more and more in their own limited trades. Indeed, there are firms where the reading even of trade journals and more general business reports is not regarded as legitimate work." (*Executive Behavior,* Strombergs, Stockholm, 1951, p. 76)

14. The concept of the manager as educator is a central thrust of Levinson, *The Exceptional Executive.* He suggests that for younger managers more attention should be given to frequent respites from daily organizational chores to get new views, to examine and digest them in work groups, and to think of their applicability to organizational problems and issues. (p. 281)

Levinson also suggests that management not press the future on the young man too hard. "If it is difficult for younger people today to make long-term commitments because change makes that seemingly less possible, then don't try to attach the bright young man to the distant future in the company." (p. 151) The implication is young people need to be motivated through their present jobs, rather than possibilities of escaping the present drag for the high-status post of the future. A similar point is made by columnist Sydney Harris. He writes: "Older people tend less to see what is in front of them than what is at a further distance; while younger persons tend to see only what is directly in front of them. Older people look into the distant past, which is in better focus for them than the immediate scene, which is blurred and confusing for it is too close-up to accommodate their angle of vision. This accounts for their impatience and incomprehension with today's activities.

"Younger people, on the other hand, never look back and rarely look forward more than a year or two; only what is right before them seems real, and both past and future seem insubstantial. And thus, their impatience and incomprehension with the 'lessons of history' and the 'dangers of the future.'" *Philadelphia Inquirer,* May 12, 1971.

15. Carlson, *op. cit.,* p. 106.

16. Holden, *et al., Top Management,* pp. 37–38. (© 1968 by McGraw-Hill, Inc. Used by permission of McGraw-Hill Book Company.)

17. *Ibid.,* p. 38.

18. In a popular book, Alvin Toffler presented a frightening view of the threats in the future; see *Future Shock* (Random House, New York, 1970). Facing this future will require courage. Friedrich Kummel writes: "The flux of time is experienced as a movement from the future to the past; life itself, however, is directed toward the future either pressing forward against the flow of time or, if capable of doing so, being carried away by the flow of time and perishing." "Time as Succession and the Problems of Duration," in Fraser, *The Voices of Time*, p. 34. The United States runs a real risk of inadequately pressing forward.

15 Conclusion—Insights, Tactics, and Implications

A multitude of issues have been considered—from the nitty-gritty of incoming telephone calls to philosophical speculation about human nature and our relation to temporal flux. Let's summerize under three categories: analytical insights, practical tactics, and philosophical implications.

ANALYTICAL INSIGHTS

(1) Achievement-oriented managers tend to be time-conscious, concerned about its use, impatient and guilty about wasting it. They want rapid feedback on their task performance. And they tend to fill up unstructured time with structured activity.

(2) Many managers work long hours with much verbal communication. Positions vary in the time devoted to responding and initiating; general executives tend to respond more to others who come to them and to be nondirective toward subordinates; service managers and operating supervisors are more authoritarian. Staff specialists and sales managers initiate the greatest proportion of activities; they are the most outwardly oriented in behavior.

(3) Managerial positions vary in the discretionary control or autonomy allowed to incumbents. Thus general executives have the most freedom to shape their behavior, functional control and sales managers less, operating supervisors and staff specialists even less, and service managers the least autonomy.

(4) Highly structured managerial positions with many explicit demands tend to be associated with directive or authoritarian leadership; more ambiguous and less time-pressured positions are associated with more diversified behavior leaning toward balanced initiation and response along with nondirective leadership.

(5) Some positions contain just too much for anyone to handle: either too much at top levels because of inability to translate organizational objectives down to operational unit goals, or too much at middle levels because of inadequately defined and limited responsibilities. As a result, stress is manifest in erratic behavior, such as withdrawal, or in excessive very short communications. In addition, too much job leads to erratic days with only short, uninterrupted periods, lack of time for internal visiting, and narrow inadequate reading.

(6) Some organizations and managers have too little to do so they have excess time. Whether due to overstaffing, inadequate delegation, or depressed sales, such a condition decreases morale and distorts behavior. Creating unnecessary problems, overreacting to normal conditions, and responding to situations that should be ignored are common developments.

(7) Fear of losing status, fear of ambiguity, and fear of losing control also waste time. Excessive and overly lengthy meetings, punishment-oriented control systems, crisis atmosphere, involvement of managers in excessive trivia, and inversion of means and ends are all symptomatic of fear.

(8) Time pressures force managers to depart from rational decision-making in a variety of ways: (a) they tend to accept the status quo and not search for "unfelt" problems, (b) because alternatives are based on past experience, they make many decisions that are conservatively biased, (c) they investigate a limited number of alternatives, and (d) they choose an acceptable rather than optimal solution. None of this is too bad if managers are aware of it. Worse, however, time

pressure encourages them to overly limit problem diagnosis in order to jump right into solutions. Managers tend to place action ahead of thought—and this is very dangerous. The more creative (and perhaps wiser) person may spend more time analyzing and less time choosing than the lesser decision maker.

(9) In making decisions, the best individuals are usually better and faster than groups; the average individual is also faster than the group, but makes more errors; the group is more accurate than most individuals, however. Younger people may be more effective in utilizing groups than older managers. Great time pressure tends to centralize groups, thus losing the advantages of diversity. Perhaps groups can be most helpful in the less time-pressured diagnostic phase of decision-making with an individual handling actual selection.

(10) In spite of the present orientation of most managers, the past does influence organizations. Through the conservative bias in decision-making and through codification of past experience in standing plans and policies, management gains necessary predictability and continuity. But it also runs the risk of inverting means and ends so that the organization loses spontaneity and direction.

(11) Although most managers might agree that they should take more time to think and plan for the future, it is very difficult to actually do so. For some, only the present has any real meaning because of their inability to delay gratification. Others may not have any faith in the future—depression, socialism, revolution, war, or other distasteful possibilities may simply discourage them from thinking about the future at all. A few competent managers may be overconfident about their ability to handle any future when it gets here. Finally, of course, present pressures may leave them inadequate time to ponder the future. When punishment for failure is close and sure, it is the dominent concern. When managers are transferred every couple of years, they may have a permanent short-timer's attitude. As a result, crisis pervades our organizations, it is difficult to train for the future, and we run a risk of being overwhelmed by a future made willy-nilly in the present.

PRACTICAL TACTICS

The following is only a partial list. Managers should be able to add to it on the basis of their experience.

(1) Segment or separate discretionary time from response time. In addition, most managers would like to reduce response time, shifting it to the discretion category. Reduction of response time may be accomplished by insulation, isolation, delegation, and simplification.

(2) Insulation involves buffering the manager from incoming communications. An assistant screens all and redirects many incoming letters, calls, and visitors. Such a tactic requires a trusted and well-qualified subordinate, for the manager runs the risk of getting out of touch with reality, and perhaps even losing control of his staff.

(3) Isolation involves physical withdrawal so that interruptions are eliminated except in emergencies. Locking the door and accepting no calls, or working in a second office are tactics. The risks are similar to those of insulation including the possibility of offending associates by not being available.

(4) Delegation does not save the organization time, but it will help the delegating manager. Guidelines for delegation include asking (a) What does not need to be done? (b) What could be handled by others? Further items include (a) retaining activities on which the manager has superiority over others; delegating others even if he is actually competent on them; (b) retaining activities concerned with change; delegating duties concerned with maintaining stability; (c) retaining discretionary activities; delegating noncontrollable response activities. Be specific on what is being delegated and do not interfere with subordinates. This requires some self-control and even courage because managers may have to buffer their subordinates from their own superiors while they are uncertain about what is going on.

(5) Simplification may save time. Grouping outgoing communications by destination and initiating them during a concentrated portion of the day is common. Some managers try to convert incoming communications into outgoing ones by not accepting them initially and having an assistant indicate they will contact the caller. Screening, grouping, and simplifying may allow the manager to handle each piece of correspondence only once; receipt and reply may be telescoped.

(6) A more reasoned choice between written and verbal communication may save time. Verbal communication is less efficient than written, yet the latter tends to be impersonal. Some guidelines: (a) Does the manager want any questions to be asked about his communication? If not, one-way announcement by letter or memo

is best. (b) Does he want the socializing that occurs with verbal communication—especially when face to face? He should be cautious about eliminating too much socializing. (c) Does he want to make himself available so the other party can raise a different subject? If yes, go talk.

(7) Discretionary time should be concentrated into good-sized chunks in order to allow uninterrupted work. In allocating this time by setting priorities, managers must distinguish between priority based on order of occurrence, urgency, and contribution to the organization. The last is the most valid criterion. Focusing on such activities requires managers to control their tendency to become consumed by the short-run, quick-feedback matters which give an illusion of progress. Managers should train themselves to view discretionary time as a resource to be invested; time requirements, therefore, should be included in the costs of alternative programs.

(8) To evaluate their allocation of time, management should draw up a list of objectives that includes results objectives (market standing, productivity, profitability, social contribution/public responsibility, innovation, and service) and process objectives (state of resources, identification, integration, collaboration, stabilization/adaption, development, and revitalization). They should be organized in order: most dominant to most deferrable. Then management should ask itself two questions: (a) Is management allocating time in accordance with its defined priorities? (b) Is management allocating sufficient time to the more deferrable, but still important objectives? Perhaps 20 percent of a manager's time should be allocated to the bottom half of the objectives list.

(9) But response time is still essential. Every manager must drift for some time responding to others; response time should not be eliminated. How much is necessary? Probably more than half of management's time. Managers must discipline themselves to understand this, and to be open, patient, and truly responsive during response time. Mentally separating response and discretionary periods may be helpful. Incoming letters, phone calls, and especially visitors are not time wasters collectively, however much they may be boring in the particular. They do not waste time unless managers become consumed by them and do not create some uninterrupted discretionary time. In short: no manager can or should manage all his time. Response time is not necessarily wasted for it can contribute to organizational

objectives, and managers should be wary of managing their time to their own interests (to minimize stress) and not in the company's interest.

(10) Try not to avoid all trivial matters. Managers should be concerned especially with issues that seem small but which are symbolic of themselves or the organization. These are matters that affect many people over a long time. Timely participation may set a tone that has great impact.

(11) See that subordinates have some discretionary time and encourage them to develop themselves. Provide some protection for promising young men against the tendency of the organization to exploit them on time-pressured, short-range issues.

(12) Fight a personal tendency to jump to solutions without adequate diagnosis. Experiment with modifying the order in which management considers alternatives; perhaps it might consider the most radical possibility first instead of last.

(13) Except where near-concensus is essential, avoid having groups handle the whole decision-making process. Utilize them to search for problems, analyze situations, and formulate alternatives, but not for actual selection.

(14) The key to planning is determination of desires. Therefore, it should begin with a reference projection of the future if management does nothing different. Then, a wishful scenario should be drafted. To do this, managers should allow themselves to dream about what they would like the organization to become. Even if the scenario is too blue-sky for implementation (and even if it is never shown to anyone), such systematic dreaming may prepare management to grasp unknown opportunity in an ambiguous future.

(15) Prospective planning attempts to fill the gap between the reference projection and the wishful scenario. It includes: (a) commitment of resources when the future is known with certainty, (b) contingency planning when the probability of various futures can be estimated, (c) responsive policy formulation when little is known about the future, and (d) proactive planning which includes active intervention in the environment. A balanced approach including all types of plans is desirable.

(16) The four-day workweek may make sense—but not in order

to provide long weekends.[1] The third day off should be on Wednesday, not Friday or Monday. And it should not be a free day; it should be a workday for education and thought. Three Wednesdays a month might be devoted to organizational development and education. Two could be reserved for individual and group consideration of the concern's future. In short, time is carved out of the present for systematic personal growth, organizational development, and most important, exploration of future possibilities, formulation of future plans, and the shaping of dreams.

PHILOSOPHICAL IMPLICATIONS

Managers have been dominated by the Newtonian view of time as constant, unvarying motion in which each interval is unique but equal. The activist ethic associated with this perspective encourages constant, short-range activity. It fosters guilt feelings when one is not active and discourages speculation. In addition, the holdover of a Victorian will to control oneself completely leads managers to view others as threats to their time use.[2] They tend to manipulate others in the interest of *their* own time utilization. But managers must be responsive part of the time because all great leaders are also servants. The paradox for managers in the United States is that they need to relax a little in order to work more effectively—relax the intensity of their work on present problems, and address themselves more to future possibilities.[3] They need to value today less and tomorrow more.

Managers tend to avoid real thought about the future because it is ambiguous. Clear-cut, programmed, short-run problems are difficult but satisfying to solve. Even planning based on extrapolation of the present is relatively concrete. In contrast, formulating wishes about the future, conceiving what we want the future to be, perceiving the future-as-history is difficult and threatening. The threat is twofold: first, because managers tend to fear the kind of time necessary for such thought; it must be unstructured, open, and seemingly ill-directed—all attributes counter to time-haunted efficiency-minded men. Wide open time like space can be frightening. Second, incorporating concern about the future into the present necessitates clarifying what we really want, and this means defining fundamental values of management, organizations, and society. Such clarification and definition is critical for all of our futures. An old American Indian proverb states, "All that is seen is temporary." Not that the present is unimportant, nor that the world is illusion, but future-oriented, time-binding managers should have a certain sense of detachment—a kind of detachment often found in men who accomplish great things while refusing to be devoured by current events and present time.

This book has raised more questions than answers. We hope that managers ask themselves some of these questions We hope that improved time management allows managers the opportunity to dream and plan their futures, their organization's, our nation's, and the world's. These dreams and plans are inseparable.

REFERENCES

1. For a complete discussion of experience with the four-day week, see Riva Poor, editor, *4 Days 40 Hours: Reporting a Revolution in Work and Leisure* (Bursk & Poor Publishing Co., Cambridge, 1970).
2. A discussion of the bankruptcy of the Victorian idea of willpower is contained in Rollo May, *Love and Will.*
3. A less aggressive attitude toward the utilization of time may even bet helpful to one's managerial career. Harry Levinson writes: "Doctors Meyer Friedman and Ray E. Roseman of Mt. Zion hospital in San Francisco . . . have followed 3182 healthy men in eleven California companies since 1960. From initial interviews they classified their research subjects into two types. Type A was characterized as aggressive, having excessive drive, ambition, and a sense of urgency. According to Dr. Rosenman, this group seemed to be fighting time itself. Type B did not have these characteristics. Type A tended to have more coronaries . . . Incidently, the type A hard driver did not usually rise to become President. Rather, the more relaxed type B did." (*Executive Stress,* p. 5)

Index